Communications
in Computer and Information Science   1340

More information about this series at http://www.springer.com/series/7899

Nicholas Olenev · Yuri Evtushenko ·
Michael Khachay · Vlasta Malkova (Eds.)

Advances in Optimization and Applications

11th International Conference, OPTIMA 2020
Moscow, Russia, September 28 – October 2, 2020
Revised Selected Papers

 Springer

Editors
Nicholas Olenev (iD)
FRC CSC RAS
Moscow, Russia

Yuri Evtushenko (iD)
FRC CSC RAS
Moscow, Russia

Michael Khachay (iD)
Krasovsky Institute of Mathematics
and Mechanics
Yekaterinburg, Russia

Vlasta Malkova (iD)
FRC CSC RAS
Moscow, Russia

ISSN 1865-0929   ISSN 1865-0937   (electronic)
Communications in Computer and Information Science
ISBN 978-3-030-65738-3   ISBN 978-3-030-65739-0   (eBook)
https://doi.org/10.1007/978-3-030-65739-0

This Springer imprint is published by the registered company Springer Nature Switzerland AG
The registered company address is: Gewerbestrasse 11, 6330 Cham, Switzerland

Preface

This book contains the second volume of the refereed proceedings of the 11th International Conference on Optimization and Applications (OPTIMA 2020)[1].

As usual, the conference was planned to be held in Montenegro on the shores of the Adriatic Sea, but the pandemic of COVID-19 disrupted our plans. Nevertheless, we could not cancel the conference, since it was dedicated to the 70th anniversary of one of its founders, the outstanding scientist, academician of the Montenegrin Academy of Sciences and Arts Prof. Milojica Jacimovic. Therefore, the Program Committee (PC) decided to organize the conference fully online, during September 30 – October 2, 2020, at the Dorodnitsyn Computing Center of the Federal Research Center "Computer Science and Control" of the Russian Academy of Sciences, Moscow, Russia.

Despite the new technical difficulties, the conference was successful. A number of vivid results were presented, many submissions passed a competitive selection and were recommended by the international PC for publication in two volumes in Springer LNCS and CCIS series, respectively. In response to the call for papers, the PC received 108 submissions from the participants. Among them, 23 full papers were selected by the PC for publication in the first proceedings volume published in Springer LNCS, vol. 12422. After the accurate revision by the authors, 18 out of the 37 remaining papers were selected by the PC for publication in this second volume. Thus, the acceptance rate for this volume was about 49%. All these papers significantly improved in response to peer reviews and discussions at the conference.

This year, the majority of the participants were still from Russia, although we are pursuing a consistent policy of increasing the number of representatives from other countries. At present, the expert community and the composition of the participants are represented by representatives of 25 countries including Argentina, Australia, Austria, Belarus, Belgium, Croatia, Finland, France, Germany, Greece, India, Israel, Italy, Kazakhstan, Montenegro, The Netherlands, Poland, Portugal, Russia, Serbia, Sweden, Taiwan, Ukraine, the UK, and the USA.

The conference was opened with the special session dedicated to the anniversary of Prof. Milojica Jacimovic, one of the founders of the conference.

The scientific program of the conference featured five plenary lectures given by outstanding scientists

- Prof. Boris T. Polyak and Ilyas Fatkhullin (Institute for Control Science, Russia), "Static feedback in linear control systems as optimization problem"
- Prof. Alexey Tret'yakov (Siedice University of Natural Sciences and Humanities, Poland), "P–regularity Theory: Applications to Optimization"
- Prof. Andrei Dmitruk (CEMI, RAS and MSU, Russia), "Lagrange Multipliers Rule for a General Extremal Problem with an Infinite Number of Constraints"

---

[1] http://www.agora.guru.ru/optima-2020.

- Prof. Nikolai Osmolovskii (Systems Research Institute, Poland), "Quadratic Optimality Conditions for Broken Extremals and Discontinuous Controls"
- Prof. Panos M. Pardalos, (University of Florida, USA), "Sustainable interdependent networks"

This year, presentations on contributed papers were partitioned into six sessions. For the session "Mathematical programming," all the papers were published in the first volume of these proceedings. Five remaining sessions are (partially) presented in this second volume. Among them are sessions on combinatorial and discrete optimization, optimal control, optimization in economics, finance, and social sciences, and global optimization; applications. The track "Combinatorial and discrete optimization" was dedicated to the memory of Prof. Alexander Kel'manov (1952–2019), who made a significant contribution to the organization of previous OPTIMA conferences.

We would like to thank all the authors for submitting to OPTIMA 2020, the members of the PC, and external reviewers for their efforts in providing exhaustive reviews. We would also like to express special gratitude to all the invited plenary speakers.

November 2020

Nicholas Olenev
Yuri Evtushenko
Michael Khachay
Vlasta Malkova

Organization

Program Committee Chairs

Milojica Jaćimović	Montenegrin Academy of Sciences and Arts, Montenegro
Yuri G. Evtushenko	Dorodnicyn Computing Centre, FRC CSC RAS, Russia
Igor G. Pospelov	Dorodnicyn Computing Centre, FRC CSC RAS, Russia
Maksat Kalimoldayev	Institute of Information and Computational Technologies, Kazakhstan

Program Committee

Majid Abbasov	St. Petersburg State University, Russia
Samir Adly	University of Limoges, France
Kamil Aida-Zade	Institute of Control Systems of ANAS, Azerbaijan
Alla Albu	Dorodnicyn Computing Centre, FRC CSC RAS, Russia
Alexander P. Afanasiev	Institute for Information Transmission Problems RAS, Russia
Yedilkhan Amirgaliyev	Süleyman Demirel University, Kazakhstan
Anatoly S. Antipin	Dorodnicyn Computing Centre, FRC CSC RAS, Russia
Sergey Astrakov	Institute of Computational Technologies, Siberian Branch of RAS, Russia
Adil Bagirov	Federation University, Australia
Evripidis Bampis	LIP6 UPMC, France
Oleg Burdakov	Linköping University, Sweden
Olga Battaïa	ISAE-SUPAERO, France
Armen Beklaryan	National Research University Higher School of Economics, Russia
Vladimir Beresnev	Sobolev Institute of Mathematics, Russia
René Van Bevern	Novosibirsk State University, Russia
Sergiy Butenko	Texas A&M University, USA
Vladimir Bushenkov	University of Évora, Portugal
Igor A. Bykadorov	Sobolev Institute of Mathematics, Russia
Alexey Chernov	Moscow Institute of Physics and Technology, Russia
Duc-Cuong Dang	INESC TEC, Portugal
Tatjana Davidovic	Mathematical Institute of Serbian Academy of Sciences and Arts, Serbia
Stephan Dempe	TU Bergakademie Freiberg, Germany

Alexandre Dolgui	IMT Atlantique, LS2N, CNRS, France
Olga Druzhinina	FRC CSC RAS, Russia
Anton Eremeev	Omsk Division of Sobolev Institute of Mathematics, Siberian Branch, RAS, Russia
Adil Erzin	Novosibirsk State University, Russia
Francisco Facchinei	Sapienza University of Rome, Italy
Vladimir Garanzha	Dorodnicyn Computing Centre, FRC CSC RAS, Russia
Alexander V. Gasnikov	Moscow Institute of Physics and Technology, Russia
Manlio Gaudioso	Università della Calabria, Italy
Alexander I. Golikov	Dorodnicyn Computing Centre, FRC CSC RAS, Russia
Alexander Yu. Gornov	Institute for System Dynamics and Control Theory, Siberian Branch, RAS, Russia
Edward Kh. Gimadi	Sobolev Institute of Mathematics, Siberian Branch, RAS, Russia
Andrei Gorchakov	Dorodnicyn Computing Centre, FRC CSC RAS, Russia
Alexander Grigoriev	Maastricht University, The Netherlands
Mikhail Gusev	N.N. Krasovskii Institute of Mathematics and Mechanics, Russia
Vladimir Jaćimović	University of Montenegro, Montenegro
Vyacheslav Kalashnikov	ITESM, Campus Monterrey, Mexico
Valeriy Kalyagin	Higher School of Economics, Russia
Igor E. Kaporin	Dorodnicyn Computing Centre, FRC CSC RAS, Russia
Alexander Kazakov	Matrosov Institute for System Dynamics and Control Theory, Siberian Branch, RAS, Russia
Mikhail Yu. Khachay	Krasovsky Institute of Mathematics and Mechanics, Russia
Oleg V. Khamisov	L. A. Melentiev Energy Systems Institute, Russia
Andrey Kibzun	Moscow Aviation Institute, Russia
Donghyun Kim	Kennesaw State University, USA
Roman Kolpakov	Moscow State University, Russia
Alexander Kononov	Sobolev Institute of Mathematics, Russia
Igor Konnov	Kazan Federal University, Russia
Vladimir Kotov	Belarus State University, Belarus
Vera Kovacevic-Vujcic	University of Belgrade, Serbia
Yury A. Kochetov	Sobolev Institute of Mathematics, Russia
Pavlo A. Krokhmal	University of Arizona, USA
Ilya Kurochkin	Institute for Information Transmission Problems, RAS, Russia
Dmitri E. Kvasov	University of Calabria, Italy
Alexander A. Lazarev	V.A. Trapeznikov Institute of Control Sciences, Russia
Vadim Levit	Ariel University, Israel
Bertrand M. T. Lin	National Chiao Tung University, Taiwan

Alexander V. Lotov	Dorodnicyn Computing Centre, FRC CSC RAS, Russia
Nikolay Lukoyanov	N.N. Krasovskii Institute of Mathematics and Mechanics, Russia
Vittorio Maniezzo	University of Bologna, Italy
Olga Masina	Yelets State University, Russia
Vladimir Mazalov	Institute of Applied Mathematical Research, Karelian Research Center, Russia
Nevena Mijajlović	University of Montenegro, Montenegro
Nenad Mladenovic	Mathematical Institute, Serbian Academy of Sciences and Arts, Serbia
Angelia Nedich	University of Illinois at Urbana Champaign, USA
Yuri Nesterov	CORE, Université Catholique de Louvain, Belgium
Yuri Nikulin	University of Turku, Finland
Evgeni Nurminski	Far Eastern Federal University, Russia
Nicholas N. Olenev	CEDIMES-Russie, Dorodnicyn Computing Centre, FRC CSC RAS, Russia
Panos Pardalos	University of Florida, USA
Alexander V. Pesterev	V.A. Trapeznikov Institute of Control Sciences, Russia
Alexander Petunin	Ural Federal University, Russia
Stefan Pickl	Bundeswehr University Munich, Germany
Boris T. Polyak	V.A. Trapeznikov Institute of Control Sciences, Russia
Yury S. Popkov	Institute for Systems Analysis, FRC CSC RAS, Russia
Leonid Popov	IMM UB RAS, Russia
Mikhail A. Posypkin	Dorodnicyn Computing Centre, FRC CSC RAS, Russia
Oleg Prokopyev	University of Pittsburgh, USA
Artem Pyatkin	Novosibirsk State University, Sobolev Institute of Mathematics, Russia
Ioan Bot Radu	University of Vienna, Austria
Soumyendu Raha	Indian Institute of Science, India
Andrei Raigorodskii	Moscow State University, Russia
Larisa Rybak	Belgorod State Technological University, Russia
Leonidas Sakalauskas	Institute of Mathematics and Informatics, Lithuania
Eugene Semenkin	Siberian State Aerospace University, Russia
Yaroslav D. Sergeyev	University of Calabria, Russia
Natalia Shakhlevich	University of Leeds, UK
Alexander A. Shananin	Moscow Institute of Physics and Technology, Russia
Angelo Sifaleras	University of Macedonia, Greece
Mathias Staudigl	Maastricht University, The Netherlands
Petro Stetsyuk	V.M. Glushkov Institute of Cybernetics, Ukraine
Alexander Strekalovskiy	Matrosov Institute for System Dynamics and Control Theory, Siberian Branch, RAS, Russia
Vitaly Strusevich	University of Greenwich, UK
Michel Thera	University of Limoges, France
Tatiana Tchemisova	University of Aveiro, Portugal

Anna Tatarczak	Maria Curie-Skłodowska University, Poland
Alexey A. Tretyakov	Dorodnicyn Computing Centre, FRC CSC RAS, Russia
Stan Uryasev	University of Florida, USA
Vladimir Voloshinov	Institute for Information Transmission Problems, RAS, Russia
Frank Werner	Otto von Guericke University Magdeburg, Germany
Oleg Zaikin	Matrosov Institute for System Dynamics and Control Theory, Siberian Branch, RAS, Russia
Vitaly G. Zhadan	Dorodnicyn Computing Centre, FRC CSC RAS, Russia
Anatoly A. Zhigljavsky	Cardiff University, UK
Julius Žilinskas	Vilnius University, Lithuania
Yakov Zinder	University of Technology, Australia
Tatiana V. Zolotova	Financial University under the Government of the Russian Federation, Russia
Vladimir I. Zubov	Dorodnicyn Computing Centre, FRC CSC RAS, Russia
Anna V. Zykina	Omsk State Technical University, Russia

Organizing Committee Chairs

Milojica Jaćimović	Montenegrin Academy of Sciences and Arts, Montenegro
Yuri G. Evtushenko	Dorodnicyn Computing Centre, FRC CSC RAS, Russia
Nicholas N. Olenev	Dorodnicyn Computing Centre, FRC CSC RAS, Russia

Organizing Committee

Gulshat Amirkhanova	Institute of Information and Computational Technologies, Kazakhstan
Natalia Burova	Dorodnicyn Computing Centre, FRC CSC RAS, Russia
Alexander Golikov	Dorodnicyn Computing Centre, FRC CSC RAS, Russia
Alexander Gornov	Institute of System Dynamics and Control Theory, Siberian Branch, RAS, Russia
Vesna Dragović	Montenegrin Academy of Sciences and Arts, Montenegro
Vladimir Jaćimović	University of Montenegro, Montenegro
Mikhail Khachay	Krasovsky Institute of Mathematics and Mechanics, Russia
Yury Kochetov	Sobolev Institute of Mathematics, Russia

Elena A. Koroleva	Dorodnicyn Computing Centre, FRC CSC RAS, Russia
Vlasta Malkova	Dorodnicyn Computing Centre, FRC CSC RAS, Russia
Nevena Mijajlović	University of Montenegro, Montenegro
Oleg Obradovic	University of Montenegro, Montenegro
Mikhail A. Posypkin	Dorodnicyn Computing Centre, FRC CSC RAS, Russia
Kirill B. Teymurazov	Dorodnicyn Computing Centre, FRC CSC RAS, Russia
Yulia Trusova	Dorodnicyn Computing Centre, FRC CSC RAS, Russia
Svetlana Vladimirova	Dorodnicyn Computing Centre, FRC CSC RAS, Russia
Victor Zakharov	FRC CSC RAS, Russia
Elena S. Zasukhina	Dorodnicyn Computing Centre, FRC CSC RAS, Russia
Ivetta Zonn	Dorodnicyn Computing Centre, FRC CSC RAS, Russia
Vladimir Zubov	Dorodnicyn Computing Centre, FRC CSC RAS, Russia

Invited Talks

Lagrange Multipliers Rule for a General Extremum Problem with an Infinite Number of Constraints

Andrei Dmitruk

CEMI RAS and MSU, Russia
https://www.researchgate.net/profile/Andrei_Dmitruk

Abstract. We consider a general optimization problem with equality and inequality constraints in a Banach space. The first is given by a level set of a nonlinear operator into another Banach space, and the latter by inclusions of images of smooth operators into closed convex sets (possibly cones) with nonempty interiors lying in some other Banach spaces. This statement covers a wide range of optimization problems both in pure mathematics and in applications. Some of its particular cases were considered earlier by many authors. We prove a first-order necessary optimality condition in the form of Lagrange multipliers rule, where the multipliers at the inequality constraints are elements of the normal cones at the corresponding points of these sets. This form is transparent for learning and convenient for application. The proof is self-contained, it uses basic facts of functional analysis and follows the line of Dubovitskii—Milyutin approach. As an application of the result, we consider an optimal control problem with state constraints, in which we obtain necessary conditions for a weak minimum.

This is joint work with Nikolai Osmolovskii.

Published in "Recent Advances of the Russian Operations Research Society" (F.Aleskerov and A.Vasin eds.), Cambridge Scholars Publishing, 2020, p. 212–232. ISBN-13: 978-1-5275-4792-6.

Quadratic Optimality Conditions for Broken Extremals and Discontinuous Controls

Nikolai Osmolovskii

Systems Research Institute, Polish Academy of Sciences, Warsaw, Poland
osmolovski@uph.edu.pl
https://www.researchgate.net/profile/Nikolai_Osmolovskii

Abstract. The talk is devoted to second-order conditions for broken extremals in variational calculus problems and for discontinuous controls in optimal control problems. A characteristic feature of the conditions under discussion is the absence of a gap between necessary and sufficient conditions. The conditions are formulated as sign-definiteness of a quadratic form on the so-called critical cone. In the first part of the talk, quadratic conditions for broken extremals are formulated in the simplest problem of the calculus of variations. In the second, we consider the optimal control problem with regular mixed constraints on the state variable and control, and the quadratic conditions for a strong local minimum are formulated for it in the case of piecewise continuous control.

Sustainable Interdependent Networks

Panos M. Pardalos

University of Florida, USA
pardalos@ufl.edu
http://www.ise.ufl.edu/pardalos/

Abstract. Sustainable interdependent networks have a wide spectrum of applications in computer science, electrical engineering, and smart infrastructures. We are going to discuss the next generation sustainability framework as well as smart cities with special emphasis on energy, communication, data analytics, and financial networks. In addition, we will discuss solutions regarding performance and security challenges of developing interdependent networks in terms of networked control systems, scalable computation platforms, and dynamic social networks.

References

Amini, M.H., Boroojeni, K.G., Iyengar, S.S., Pardalos, P., Blaabjerg, F., Madni, A.M. (eds.) Sustainable Interdependent Networks: From Theory to Application. Springer, Cham (2018). https://doi.org/10.1007/978-3-319-74412-4

Amini, M.H., Boroojeni, K.G., Iyengar, S.S., Pardalos, P., Blaabjerg, F., Madni, A.M. (eds.) Sustainable Interdependent Networks: From Smart Power Grids to Intelligent Transportation Networks. Springer, Cham (2019). https://doi.org/10.1007/978-3-319-98923-5

Rassia, S.Th., Pardalos, P.M. (eds.) Smart City Networks: Through the Internet of Things. Springer, Cham (2017). https://doi.org/10.1007/978-3-319-61313-0

Kalyagin, V.A., Pardalos, P.M., Rassias, T.M. (eds.) Network Models in Economics and Finance. Springer, Cham (2014). https://doi.org/10.1007/978-3-319-09683-4

Carpi, L., et al.: Assessing diversity in multiplex networks. Nat. Sci. Rep. (2019)

Schieber, T.A., Carpi, L., Díaz-Guilera, A., Pardalos, P.M., Masoller, C., Ravetti, M.G.: Quantification of network structural dissimilarities. Nat. Commun. **8** (2017). Article number: 13928

Static Feedback in Linear Control Systems as Optimization Problem

Boris T. Polyak

Institute for Control Science, Moscow, Russia
boris@ipu.ru
https://www.researchgate.net/profile/Boris_Polyak2

Abstract. The linear quadratic regulator is the fundamental problem of optimal control. Its state feedback version was set and solved in the early 1960s. However, the static output feedback problem has no explicit-form solution. It is suggested to look at both of them from another point of view as a matrix optimization problem, where the variable is a feedback matrix gain. The properties of such a function are investigated, it turns out to be non-convex, with the possible non-connected domain. Moreover, it is not L-smooth on the entire domain but has this property on sublevel sets. Nevertheless, a specially adopted gradient method for its minimization converges to the optimal solution in the state feedback case and to a stationary point in the output feedback case. The results can be extended for the general framework of the reduced gradient method for optimization with equality-type constraints. Directions for future research are addressed.

This is joint work with Ilyas Fatkhullin.

P-regularity Theory: Applications to Optimization

Alexey Tret'yakov[1,2,3]

[1] System Research Institute, Polish Academy of Sciences, Poland
[2] Siedlce University, Faculty of Sciences, 08-110 Siedlce, Poland,
[3] Dorodnicyn Computing Centre, FRC CSC RAS, Russia
tret@ap.siedlce.pl
https://www.researchgate.net/profile/Alexey_Tretyakov

Abstract. We present recent advances in the analysis of nonlinear structures and their applications to nonlinear optimization problems with constraints given by nonregular mappings or other singularities obtained within the framework of the p-regularity theory developed over the last 20 years. In particular, we address the problem of description of the tangent cone to the solution set of the operator equation, optimality conditions, and solution methods for optimization problems.

This is joint work with Yuri Evtushenko and Vlasta Malkova.

Contents

Global Optimization

Global Optimization Method with Numerically Calculated Function Derivatives

Victor Gergel[ID] and Alexander Sysoyev[✉][ID]

Lobachevsky State University of Nizhny Novgorod,
Nizhny Novgorod, Russian Federation
gergel@unn.ru, alexander.sysoyev@itmm.unn.ru

Abstract. The paper proposes a method for solving computationally time-consuming multidimensional global optimization problems. The developed method combines the use of a nested dimensional reduction scheme and numerical estimates of the objective function derivatives. Derivatives significantly reduce the cost of solving global optimization problems, however, the use of a nested scheme can lead to the fact that the derivatives of the reduced function become discontinuous. Typical global optimization methods are highly dependent on the continuity of the objective function. Thus, to use derivatives in combination with a nested scheme, an optimization method is required that can work with discontinuous functions. The paper discusses the corresponding method, as well as the results of numerical experiments in which such an optimization scheme is compared with other known methods.

Keywords: Multidimensional optimization · Global search algorithms · Lipschitz condition · Numerical estimations of derivative values · Dimensionality reduction · Discontinuous functions · Numerical experiments

1 Introduction

The global (or multiextremal) unconstrained optimization problem [5,6,15,17, 19–21,28,30–32] can be stated as follows

$$\varphi(y^*) = \min\{\varphi(y) \colon y \in D\}, \tag{1}$$

where search domain D represents an N-dimensional hyperinterval:

$$D = \{y \in \mathbb{R}^N \colon a_i \le y_i \le b_i, \ i = \overline{1, N}\}.$$

The objective function $\varphi(y)$ is assumed to be a multiextremal one. Also one of the commonly used assumptions is that the minimized function satisfies the Lipschitz condition

$$|\varphi(y_2) - \varphi(y_1)| \le L\|y_2 - y_1\|, \ y_1, y_2 \in D. \tag{2}$$

Supported by Russian Foundation for Basic Research (grant 19-07-00242).

N. Olenev et al. (Eds.): OPTIMA 2020, CCIS 1340, pp. 3–14, 2020.
https://doi.org/10.1007/978-3-030-65739-0_1

where $L > 0$ is the Lipschitz constant, and $\| \cdot \|$ denotes the Euclidean norm in $\mathbb{R}^N$.

The Lipschitz condition corresponds to the assumption of a limited variation of the function value at limited variations of its parameters. This condition allows making the estimates of potential behaviour of the function $\varphi(y)$ based on a finite set of its values computed at some points in the search domain D.

To solve problem (1) numerically, optimization methods usually generate a sequence of points y_k, which converges to the global optimum y^*. The amount of computation can grow exponentially with an increase in the number of variable parameters N.

One approach to reduce computational costs is to use differentiability of the objective function. In this case the fulfilment of the Lipschitz condition (2) may be expanded onto the partial derivatives $\varphi_i'(y), 1 \leq i \leq N$ of the objective function as well i.e.

$$|\varphi_i'(y_2) - \varphi_i'(y_1)| \leq L_i\|y_2 - y_1\|, \ y_1, y_2 \in D, 1 \leq i \leq N, \tag{3}$$

where $L_i > 0, 1 \leq i \leq N$ are the corresponding Lipschitz constants for the partial derivatives $\varphi_i'(y), 1 \leq i \leq N$ [1,2,8–10,16,23,24,26].

However, in some applied optimization problems the computing of the derivatives may be restricted or even impossible. In this case the usage of the global optimization methods, in which the necessary values of derivatives are computed numerically may be useful [10,12,13,18].

A widely used approach to solve the problem of multidimensional optimization is to reduce it to one-dimensional. This reduction may be based on using Peano (space-filling) curves [22,30], the nested multistep reduction scheme [4,25], the diagonal generalization technique [9,21,24], etc. Thus, one-dimensional optimization algorithms can be effectively applied in the multidimensional case [1,2,4,7,8,10]. It is known that using the nested multistep reduction scheme can lead for executing some redundant global search iterations [4,25,30]. This deficiency can be diminished by using the values of derivatives of the objective function – see the results of numerical experiments given in Sect. 5.

In this paper, the global optimization algorithm utilizing the numerical derivatives of the objective function $\varphi(y)$ is considered. In Sect. 2, the base one-dimensional algorithm utilizing the numerical derivatives is given. Section 3 introduces a nested dimension reduction scheme that allows one to generalize the proposed one-dimensional algorithm to solving multidimensional global optimization problems. Section 4 describes an approach of usage derivatives in combination with a nested scheme, that can work with discontinuous functions. In Sect. 5, the results of the numerical experiments are considered, which confirm the developed approach to be promising.

2 One-Dimensional Global Optimization Algorithm Utilizing Numerical Derivatives

The proposed optimization algorithm is based on the adaptive global method using derivatives (AGMD) [8,9], designed to solve one-dimensional global optimization problems

$$\varphi(x^*) = \min\{\varphi(x) \colon x \in [a, b]\}. \tag{4}$$

The adaptive global method using numerical derivatives (AGMND) is a modification of AGMD, in which the values of the first derivative of the objective function are replaced by their numerical estimates [9,12].

Consider the computational scheme of the AGMND. The first two iterations are performed at the boundary points a and b. Then let $k, k > 1$ iterations of the global search were completed, and the values of the objective function $\varphi(x)$ have been computed at each iteration (hereinafter, these computations will be called *trials*). The test point of the next $(k + 1)$ optimization iteration is determined by the following rules.

Rule 1. Renumber the points of previous trials by subscripts in increasing order

$$a = x_0 < x_1 < \ldots < x_i < \ldots < x_k = b. \tag{5}$$

Rule 2. Compute the numerical estimations of the first derivatives of $\varphi(x)$ at the points of the executed search iterations $x_i, 0 \le i \le k$ according to the expressions:

$$\dot{z}_i = \begin{cases} \frac{z_{i+1} - z_i}{x_{i+1} - x_i}, i = 0, \\ \frac{z_i - z_{i-1}}{x_i - x_{i-1}}, 1 \le i \le k, \end{cases} \tag{6}$$

hereinafter $z_i, 0 \le i \le k$ denotes $\varphi(x_i)$.

Rule 3. Compute the estimation of the Lipschitz constant from (3) for the first derivative of the optimized function

$$m = \begin{cases} rM, M > 0, \\ 1, \quad M = 0, \end{cases} \tag{7}$$

where

$$M = \max(M_i), 1 \le i \le k, \tag{8}$$

$$M_i = \max \begin{cases} |\dot{z}_i - \dot{z}_{i-1}|/|x_i - x_{i-1}|, \\ -2[z_i - z_{i-1} - \dot{z}_{i-1}(x_i - x_{i-1})]/(x_i - x_{i-1})^2|, \\ 2[z_i - z_{i-1} - \dot{z}_i(x_i - x_{i-1})]/(x_i - x_{i-1})^2|, \end{cases} \tag{9}$$

and $r > 1$ is the reliability parameter of the algorithm.

Rule 4. Compute the characteristic $R(i)$ for each interval $(x_{i-1}, x_i), 1 \le i \le k$ according to the following expressions to estimate the minimum possible values of $\varphi(x)$ in the interval $((x_{i-1}, x_i)$

$$R(i) = \begin{cases} \widehat{\varphi}_i(\widehat{x}_i), & \widehat{x}_i \in [\bar{x}_i, \bar{\bar{x}}_i], \\ \min(\widehat{\varphi}_i(\bar{x}_i), \widehat{\varphi}_i(\bar{\bar{x}}_i)), & \widehat{x}_i \notin [\bar{x}_i, \bar{\bar{x}}_i], \end{cases} \tag{10}$$

where
$$\widehat{x}_i = \frac{-\dot{z}_{i-1} + m(\bar{x}_i - x_{i-1}) + mx_i}{m}, \tag{11}$$

and the auxiliary functions (minorants) $\widehat{\varphi}_i(x), 1 \leq i \leq k$ take the form

$$\widehat{\varphi}_i(x) = \begin{cases} \widehat{\varphi}_{i1}(x) = z_{i-1} + \dot{z}_{i-1}(x_i - x_{i-1}) - 0.5m(x - x_{i-1})^2, & x \in (x_{i-1}, \bar{x}_i) \\ \widehat{\varphi}_{i2}(x) = A_i(x - \bar{x}_i) + 0.5m(x - \bar{x}_i)^2 + B_i, & x \in [\bar{x}_i, \bar{\bar{x}}_i], \\ \widehat{\varphi}_{i3}(x) = z_i - \dot{z}_i(x - x_i) - 0.5m(x - x_i)^2, & x \in (\bar{\bar{x}}_i, x_i], \end{cases} \tag{12}$$

where

$$\begin{aligned} A_i &= \dot{z}_{i-1} - m(\bar{x}_i - x_{i-1}), \\ B_i &= \widehat{\varphi}_{i1}(\bar{x}_i), \\ \bar{x}_i &= \frac{(z_{i-1} - \dot{z}_{i-1}x_{i-1}) - (z_i - \dot{z}_ix_i) + m(x_i^2 - x_{i-1}^2)/2 - md_i^2}{m(x_i - x_{i-1}) + (\dot{z}_i - \dot{z}_{i-1})} \\ \bar{\bar{x}}_i &= \frac{(z_{i-1} - \dot{z}_{i-1}x_{i-1}) - (z_i - \dot{z}_ix_i) + m(x_i^2 - x_{i-1}^2)/2 + md_i^2}{m(x_i - x_{i-1}) + (\dot{z}_i - \dot{z}_{i-1})} \\ d_i &= (x_i - x_{i-1})/2 - (\dot{z}_i - \dot{z}_{i-1})/2m. \end{aligned} \tag{13}$$

Each characteristic $R(i), 1 \leq i \leq k$ calculated in this way is an estimation of the minimum possible value of the minorant $\widehat{\varphi}_i(x)$ from (12) in the intervals $[x_{i-1}, x_i]$ and the estimation of the minimum possible values of $\varphi(x)$ in these intervals.

Rule 5. Find the interval (x_{t-1}, x_t) with the minimal characteristic $R(t)$

$$R(t) = \min\{R(i) : 1 \leq i \leq k\}. \tag{14}$$

In the case when there are several intervals satisfying (14), for definiteness, the interval with the minimum number t is taken.

Rule 6. Compute the next point of the next trial x^{k+1} accordingly

$$x^{k+1} = \begin{cases} \widehat{x}_t, & \widehat{x}_t \in [\bar{x}_t, \bar{\bar{x}}_t], \\ \bar{x}_t, & \widehat{\varphi}(\bar{x}_t) \leq \widehat{\varphi}(\bar{\bar{x}}_t), \\ \bar{\bar{x}}_t, & \widehat{\varphi}(\bar{x}_t) > \widehat{\varphi}(\bar{\bar{x}}_t). \end{cases} \tag{15}$$

The stopping condition is defined by the following relation

$$|x_t - x_{t-1}| \leq \varepsilon, \tag{16}$$

where ε is the accuracy, $\varepsilon > 0$. The minimum computed value of the objective function is accepted as the current estimate of the global minimum value i.e:

$$\varphi^* = \min\{z_i : 0 \leq i \leq k\}. \tag{17}$$

Note 1. The computing of the numerical estimations $\dot{z}_i, 0 \leq i \leq k$ of the first derivative of the function $\varphi(x)$ can be performed also using the three-point approximating expressions:

$$\dot{z}_i = \begin{cases} \frac{1}{H_1^2}\left(-(2+\delta_2)z_0 + \frac{(1+\delta_2)^2}{\delta_2}z_1 - \frac{1}{\delta_2}z_2\right), & i = 0, \\ \frac{1}{H_i^{i+1}}\left(-\delta_{i+1}z_{i-1} + \frac{\delta_{i+1}^2-1}{\delta_{i+1}}z_i + \frac{1}{\delta_{i+1}}z_{i+1}\right), & 1 < i < k, \\ \frac{1}{H_{k-1}^k}\left(\delta_k z_{k-2} - \frac{(1+\delta_k)^2}{\delta_k}z_{k-1} + \frac{(2\delta_k+1)}{\delta_k}z_k\right), & i = k, \end{cases} \tag{18}$$

where $H_i^{i+1} = h_i + h_{i+1}, \delta_{i+1} = \frac{h_{i+1}}{h_i}$ and $h_i = x_i - x_{i-1}$, see [13]. These formula used three points of trials, and can be used in the **Rule 2** if $k > 2$.

Note 2. For the applicability of the computational scheme described above, the fulfilment of the following inequalities

$$x_{i-1} < \bar{x}_i < \bar{\bar{x}}_i < x_i \tag{19}$$

for all intervals $(x_{i-1}, x_i), 1 \leq i \leq k$ is necessary. If the estimate of the Lipschitz constant m computed in (7) is insufficient and the condition (19) is violated for some $1 \leq i \leq k$ the value m should be refined. Thus, the maximum root of the equations

$$\begin{cases} -(z_i - z_{i-1}) + 0.5(\dot{z}_i + \dot{z}_{i-1}) + 0.25m(x_i - x_{i-1})^2 - \frac{(\dot{z}_i - \dot{z}_{i-1})^2}{4m} = 0, \\ (z_i - z_{i-1}) - 0.5(\dot{z}_i + \dot{z}_{i-1}) + 0.25m(x_i - x_{i-1})^2 - \frac{(\dot{z}_i - \dot{z}_{i-1})^2}{4m} = 0, \end{cases} \tag{20}$$

was selected as m in this case for AGMD [8].

3   Nested Dimensionality Reduction Scheme

One approach to solve multidimensional optimization problems is dimension reduction, which allows to use effective one-dimensional optimization methods. In this paper, dimension reduction is performed using the well-known nested scheme [4, 25, 28, 30, 31]. According to this scheme, the solving of a multidimensional optimization problem (1) can be obtained by solving a series of nested one-dimensional problems:

$$\min\{\varphi(y) : y \in D\} = \min_{[a_1,b_1]} \ldots \min_{[a_N,b_N]} \varphi(y_1, \ldots, y_N). \tag{21}$$

In other words the solving of problem (1) is reduced to solving a one-dimensional problem:

$$\varphi(y^*) = \min_{y \in D} = \min_{y_1 \in [a_1,b_1]} \widetilde{\varphi}_1(y_1), \tag{22}$$

where

$$\widetilde{\varphi}_i(y_i) = \varphi_i(y_1, \ldots, y_i) = \min_{y_{i+1} \in [a_{i+1},b_{i+1}]} \varphi_i(y_1, \ldots, y_i, y_{i+1}), 1 \leq i \leq N, \tag{23}$$

$$\widetilde{\varphi}_N(y_1, \ldots, y_N) = \varphi(y_1, \ldots, y_N).$$

The one-dimensional function in (22) is constructed according to a general recursive scheme – in order to compute the values $\widetilde{\varphi}_1(y_1)$ for some given value of the variable $y_1 = \widehat{y}_1$ it is necessary to minimize the function

$$\widetilde{\varphi}_2(y_2) = \varphi_2(\widehat{y}_1, y_2). \tag{24}$$

With respect to y_2 the function $\widetilde{\varphi}_2(y_2)$ is a one-dimensional one as well since the value of the variable y_1 is given and fixed one. Next, in turn, in order to compute the value of $\widetilde{\varphi}_2(y_2)$ at the point $y_2 = \widehat{y}_2$, it is necessary to minimize the function

$$\widetilde{\varphi}_3(y_3) = \varphi_3(\widehat{y}_1, \widehat{y}_2, y_3), \tag{25}$$

and so forth.

Additional information on the nested dimensionality reduction scheme and its applications for solving the multidimensional global optimization problems can be found, for example, in [4,25,30,31].

4 Global Optimization Algorithm for Discontinuous Functions

To solve problem (1), the method with derivatives can only be used if the objective function is smooth. But one-dimensional functions $\widetilde{\varphi}_i(y_i), 1 \leq i < N$ from (23) (except the function of the last decomposition level $\widetilde{\varphi}_N(y_N)$) in the nested reduction scheme can be non-smooth at some points i.e. the derivatives of these functions can be discontinuous at these points.

Strictly speaking, the AGMND method described in Sect. 2 can only be used to minimize one-dimensional function $\widetilde{\varphi}_N(y_N)$ at the last level of decomposition. The paper [11] presents the results of experiments comparing a number of optimization methods, which showed that the AGMND method provides good efficiency even in the case of nonsmooth reduced functions. But with increasing the problem dimensionality (N) the number of discontinuity points of the derivatives can grow exponentially and increase the number of executed trials.

To handle the discontinuity problem one can use the combined method which uses AGMND to minimize the one-dimensional function $\widetilde{\varphi}_N(y_N)$) and any one-dimensional method without derivatives for the remaining functions $\widetilde{\varphi}_i(y_i), 1 \leq i < N$.

Another way is to use an optimization method that can work with discontinuous functions. As such a method, the paper considers the modified Strongin algorithm [29]. Consider its computational scheme.

The first iteration is performed at any point in the interval (a, b). Then let $k, k > 1$ iterations of the global search were completed. The test point of the next $(k + 1)$ optimization iteration is determined by the following rules.

Rule 1. Renumber the points of previous trials by subscripts in increasing order

$$a = x_0 < x_1 < \ldots < x_i < \ldots < x_k = b. \tag{26}$$

Rule 2. Compute

$$\mu_i = \frac{|z_i - z_{i-1}|}{|x_i - x_{i-1}|}, 1 \leq i \leq k. \tag{27}$$

Rule 3. Reorder μ_i values in descending order

$$\mu(1) > \mu(2) > \ldots > \mu(k) \tag{28}$$

and determine a minimal number p such that

$$\frac{\mu(p)}{\mu(p+1)} \geq Q, 1 \leq p < qk, \tag{29}$$

where q and Q are given numbers $(0 < q < 1 < Q)$.

Construct a subset of numbers

$$J = \{i : 1 \leq i \leq k, \mu_i = \mu(j), 1 \leq j \leq p\} \tag{30}$$

of those intervals to which sufficiently large μ_i correspond.

Rule 4. Set for all intervals

$$\delta_i = \begin{cases} sign(z_i - z_{i-1}), & i \in J, \\ 0, & i \notin J, \end{cases} 1 \leq i \leq k. \tag{31}$$

Rule 5. Compute the estimation of the Lipschitz constant

$$\mu = \max(\mu_i), 1 \leq i \leq k \tag{32}$$

and the characteristic $R(i)$ for each interval $(x_{i-1}, x_i), 1 \leq i \leq k$

$$R(i) = (1 + |\delta_i|)\Delta_i + (1 - |\delta_i|)\frac{(z_i - z_{i-1})^2}{(r\mu)^2 \Delta_i} - 2\frac{(1 - \delta_i)z_i + (1 + \delta_i)z_{i-1}}{r\mu}, \tag{33}$$

where $\Delta_i = x_i - x_{i-1}$.

Rule 6. Find the interval (x_{t-1}, x_t) with the maximal characteristic $R(t)$

$$R(t) = \max\{R(i) : 1 \leq i \leq k\}. \tag{34}$$

Rule 7. Compute the next point of the next trial x^{k+1} accordingly

$$x^{k+1} = \frac{x_t + x_{t-1}}{2} - (1 - |\delta_t|)\frac{z_t - z_{t-1}}{2r\mu}. \tag{35}$$

5 Results of Numerical Experiments

The first experiment was performed on a series of 20 one-dimensional test global optimization problems accumulated in [14]. In this experiment we compared the following methods: the Galperin Algorithm (GA) [7], the Piyavskii-Shubert Algorithm (PA) [22,27], the Strongin Algorithm (SA) [28,30], the Brent Algorithm (BA) [3], AGMD, AGMND and Discontinuous modification of the Strongin Algorithm (DSA) described in Sect. 4.

Table 1. The results of comparison of one-dimensional methods of global optimization

	GA	PA	SA	BA	AGMD	AGMND	DSA
1	377	149	127	43	27	16	80
2	308	155	135	24	27	12	45
3	581	195	224	153	98	59	82
4	923	413	379	16	27	11	69
5	326	151	126	45	23	17	26
6	263	129	112	123	39	25	521
7	383	153	115	23	25	12	39
8	530	185	188	148	88	45	88
9	314	119	125	44	26	13	97
10	416	203	157	27	25	10	66
11	779	373	405	47	41	26	142
12	746	327	271	30	37	21	38
13	1829	993	472	69	89	19	93
14	290	145	108	34	30	15	31
15	1613	629	471	50	47	29	96
16	992	497	557	109	75	23	304
17	1412	549	470	124	65	12	134
18	620	303	243	8	21	10	84
19	302	131	117	21	21	13	47
20	1412	493	81	99	32	25	277
Average	780.80	314.60	244.15	61.85	43.15	20.65	117.95

The purpose of this experiment is to determine whether it is possible to use the DSA method to solve problems in which the objective function is smooth. Numerical results are presented in Table 1, which shows the number of iterations performed by each algorithm to solve test optimization problems with a given accuracy.

Analyzing the results, it should be taken into account that all the considered methods at each iteration calculate the value of the objective function, except for the AGMD method, in which the value of the first derivative is calculated.

Table 1 shows the significant superiority of the AGMND method over all others, including the BA and AGMD methods. We also see that the DSA method is in fourth place in terms of the average number of iterations, and in problems 20 and 6 it takes fifth and seventh positions, respectively. So the numerical results show that it is inefficient to use a method that takes into account possible discontinuities of the function to solve one-dimensional global optimization problems with smooth objective functions.

The second experiment was performed on a series of 100 well-known two-dimensional multiextremal test functions [28, 30, 31] defined by the relations:

$$\varphi(y_1, y_2) = -\left\{ \left(\sum_{i=1}^{7} \sum_{j=1}^{7} [A_{ij} a_{ij}(y_1, y_2) + B_{ij} b_{ij}(y_1, y_2)] \right)^2 \right.$$

$$\left. + \left(\sum_{i=1}^{7} \sum_{j=1}^{7} [C_{ij} a_{ij}(y_1, y_2) + D_{ij} b_{ij}(y_1, y_2)] \right)^2 \right\}^{\frac{1}{2}}, \tag{36}$$

$$a_{ij}(y_1, y_2) = \sin(\pi i y_1) \sin(\pi j y_2),$$
$$b_{ij}(y_1, y_2) = \cos(\pi i y_1) \cos(\pi j y_2),$$

where $0 \leq y_1, y_2 \leq 1$, were used. The values $-1 \leq A_{ij}, B_{ij}, C_{ij}, D_{ij} \leq 1$ are the independent random generated parameters distributed uniformly over the interval $[-1, 1]$.

Taking into account the presented results, in this experiment only algorithms with the best performance were used (namely, SA, BA, AGMD, AGMND together with the nested dimension reduction scheme). Also we evaluated the efficiency of three combined methods: SA-D, SA-ND, DSA-ND. In these methods SA (or DSA) method was applied to optimize the reduced one-dimensional functions $\widetilde{\varphi}_1(y_1)$ from (23) and AGMD or AGMND method was used to optimize the functions $\widetilde{\varphi}_2(y_2)$. Additionally, the experiment involved a version of the SA algorithm with dimension reduction using Peano space filling curves (SA-P).

In all SA-based methods, an adaptive scheme of computing the reliability parameter r from (7) was applied: $r = 3 + 10/k$ where k is the number of executed trials. The accuracy of solving the optimization problems was $\varepsilon = 0.001$.

The average number of iterations executed by each method until the stopping condition is satisfied when solving 100 test problems (36) is presented in Table 2.

Table 2. Average number of executed iterations when solving 100 test problems (36)

	SA	BA	AGMD	AGMND	SA-D	SA-ND	DSA-ND	SA-P
Average	1974.75	2626.31	824.18	494.74	924.86	754.34	1022.48	696.69

These results demonstrate that the AGMND method shows the best performance (the smallest number of trials). The next most efficient method is SA-P. The combined SA-ND and AGMD methods show similar characteristics with a slight advantage of SA-ND. BA effectiveness is the smallest. Again, for the correct analysis of the results obtained, it should be kept in mind that in the AGMD method, each trial includes the calculation of derivatives, while in the GSA, BA and AGMND methods only objective functions are calculated.

Also we can see that the effectiveness of the DSA-ND method is lower than all others except for SA and BA methods.

For a more detailed comparison of the effectiveness of the AGMD and SA-ND methods, the operational characteristics were constructed according to the experimental results. The operational characteristic is a graph of the number of solved problems (the ordinate axis) vs the number of executed trials (the abscissa axis) [28, 30, 31]. The operational characteristics of the compared methods are presented in Fig. 1.

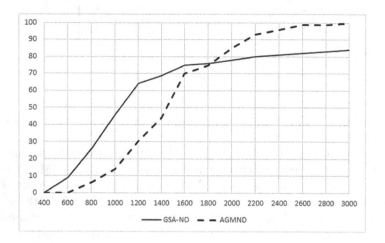

Fig. 1. Operational characteristics of the compared optimization methods. The vertical axis is the percentage of problems solved with the required accuracy, the horizontal axis is the number of executed trials

As you can see, DSA-ND solved 75% problems faster than AGMND method, but later reached 100% solvability. The comparison results show that the AGMND method demonstrates higher efficiency in solving multidimensional problems than the DSA-ND, even taking into account the possible discontinuity of the reduced functions $\widetilde{\varphi}_1(y_1)$ from (23).

6   Conclusion

In this paper, an efficient approach for solving computationally time-consuming multidimensional global optimization problems is proposed.

The developed method combines the use of a nested dimensional reduction scheme and numerical estimates of the objective function derivatives. As known derivatives significantly reduce the cost of solving global optimization problems. However, the use of a nested scheme can lead to the fact that the derivatives of the reduced function become discontinuous. Thus, to use derivatives in combination with a nested scheme, we can choose one of three possible approaches: 1) use the method with derivatives despite the possible non-smoothness of the reduced functions, 2) use methods that combine the use of derivatives to minimize the one-dimensional function $\widetilde{\varphi}_N(y_N)$) and any one-dimensional method without derivatives for other functions $\widetilde{\varphi}_i(y_i), 1 \leq i < N$ from (23) , 3) use a method that can work with discontinuous functions.

The paper discusses the methods corresponding to all the indicated approaches, presents the results of numerical experiments comparing the effectiveness of the selected methods on one-dimensional and two-dimensional optimization problems.

References

1. Baritompa, W.: Accelerations for a variety of global optimization methods. J. Global Optim. **4**, 37–45 (1994)
2. Breiman, L., Cutler, A.: A deterministic algorithm for global optimization. Math. Program. **58**, 179–199 (1993)
3. Brent, R.P.: Algorithms for Minimization Without Derivatives. Prentice-Hall, Englewood Cliffs (1973)
4. Dam, E.R., Husslage, B., Hertog, D.: One-dimensional nested maximin designs. J. Glob. Optim. **46**, 287–306 (2010)
5. Floudas, C.A., Pardalos, M.P.: State of the Art in Global Optimization. Computational Methods and Applications. Kluwer Academic Publishers, Dordrecht (1996)
6. Floudas, C.A., Pardalos, M.P.: Recent Advances in Global Optimization. Princeton University Press, Princeton (2016)
7. Galperin, E.A.: The cubic algorithm. J. Math. Anal. Appl. **112**, 635–640 (1985)
8. Gergel, V.P.: A method of using derivatives in the minimization of multiextremum functions. Comput. Math. Math. Phys. **36**, 729–742 (1996). (In Russian)
9. Gergel, V.P.: A global optimization algorithm for multivariate function with Lipschitzian first derivatives. J. Glob. Optim. **10**, 257–281 (1997)
10. Gergel, V., Goryachih, A.: Global optimization using numerical approximations of derivatives. In: Battiti, R., Kvasov, D.E., Sergeyev, Y.D. (eds.) LION 2017. LNCS, vol. 10556, pp. 320–325. Springer, Cham (2017). https://doi.org/10.1007/978-3-319-69404-7_25
11. Gergel, V., Goryachih, A.: Multidimensional global optimization using numerical estimates of objective function derivatives. In: Optimization Methods and Software (2019)
12. Goryachih, A.S., Rachinskaya, M.A.: Multidimensional global optimization method using numerically calculated derivatives. Proc. Comput. Sci. **119**, 90–96 (2017)
13. Griewank, A., Walther, A.: Evaluating Derivatives: Principles and Techniques of Algorithmic Differentiation. SIAM (2008)
14. Hansen, P., Jaumard, B., Lu, S.H.: Global optimization of univariate Lipshitz functions. II. New algorithms and computational comparison. Math. Program. **55**, 273–292 (1992)
15. Horst, R., Tuy, H.: Global Optimization: Deterministic Approaches. Springer, Heidelberg (1990). https://doi.org/10.1007/978-3-662-02598-7
16. Lera, D., Sergeyev, Y.D.: Acceleration of univariate global optimization algorithms working with Lipschitz functions and Lipschitz first derivatives. SIAM J. Optim. **23**, 508–529 (2013)
17. Locatelli, M., Schoen, F.: Global Optimization: Theory, Algorithms, and Applications. SIAM (2013)
18. Nocedal, J., Wright, S.: Numerical Optimization. Springer, Heidelberg (2006). https://doi.org/10.1007/978-0-387-40065-5

19. Pardalos, M.P., Zhigljavsky, A.A., Žilinskas, J.: Advances in Stochastic and Deterministic Global Optimization. Springer, Heidelberg (2016). https://doi.org/10.1007/978-3-319-29975-4
20. Paulavičius R., Žilinskas J.: Simplicial Global Optimization. Springer Briefs in Optimization. Springer, Heidelberg (2014). https://doi.org/10.1007/978-1-4614-9093-7
21. Pintér, J.D.: Global Optimization in Action (Continuous and Lipschitz Optimization: Algorithms, Implementations and Applications). Kluwer Academic Publishers, Dordrecht (1996)
22. Piyavskij, S.: An algorithm for finding the absolute extremum of a function. Computat. Math. Math. Phys. **12**, 57–67 (1972). (In Russian)
23. Sergeyev, Y.D.: Global one-dimensional optimization using smooth auxiliary functions. Math. Program. **81**, 127–146 (1998)
24. Sergeyev, Y.D.: A deterministic global optimization using smooth diagonal auxiliary functions. Commun. Nonlinear Sci. Numer. Simul. **21**, 99–111 (2015)
25. Shi, L., Ólafsson, S.: Nested partitions method for global optimization. Oper. Res. **48**, 390–407 (2000)
26. Shpak, A.: Global optimization in one-dimensional case using analytically defined derivatives of objective function. Comput. Sci. J. Mold. **3**, 168–184 (1995)
27. Shubert, B.O.: A sequential method seeking the global maximum of a function. SIAM J. Numer. Anal. **9**, 379–388 (1972)
28. Strongin, R.G.: Numerical Methods in the Multiextremal Problems (Information-Statistical Algorithms). Nauka (1978). (In Russian)
29. Strongin R.G.: Search of global optimum. Znanie (1990). (In Russian)
30. Strongin, R.G., Sergeyev, Ya.D.: Global Optimization with Non-convex Constraints: Sequential and Parallel Algorithms. Kluwer Academic Publishers, Dordrecht (2000). 2nd edn. 2013, 3rd edn. 2014
31. Strongin, R.G., Gergel, V.P., Grishagin, V.A., Barkalov K.A.: Parallel Computations in the Global Optimization Problems. MSU Publishing (2013). (In Russian)
32. Zhigljavsky, A., Žilinskas, A.: Stochastic Global Optimization. Springer, Berlin (2008). https://doi.org/10.1007/978-0-387-74740-8

Improving of the Identification Algorithm for a Quasilinear Recurrence Equation

Anatoly V. Panyukov[✉][iD] and Yasir Ali Mezaal

Federal State Autonomous Educational Institution of Higher Education,
South Ural State University (National Research University),
76, Lenin Prospekt, 454080 Chelyabinsk, Russia
info@susu.ru
http://www.susu.ru/en

Abstract. Identification of quasilinear recurrence equations (QRE) may be reduced to the problem of regression analysis with mutually dependent observable variables. It is possible to use the generalized least deviations method (GLDM) for such problems. GLDM-estimation consists of solving the sequence of the WLDM-estimation problems. We propose the algorithm to solve the WLDM-estimation problem. Computational complexity of this algorithm does not exceed the quantity $O(N^2 T + T^2)$, where N is the number of coefficients in the considered equation, T is the number of observed readings. The computational complexity of solving practical GLDM estimation problems does not exceed $O(N^3 T + NT^2)$. Results of computational experiments to solve the problem of identifying the recurrence equation of the stock market index in Iraq by original data from the site "ISX-IQ.net" are presented. This results show the possibility to apply a second order quasilinear recurrence equation with quadratic nonlinearity for these purposes. Perhaps increasing the order of the recurrence equation and the accuracy of the calculations give better results.

Keywords: Least deviation method · Autoregressive model · Linear programming · Gradient projection method · Computational complexity

1 Introduction

We consider the problem to determine the coefficients

$$a_1, a_2, a_3 \ldots, a_m, b_1, b_2, b_3 \ldots, b_n \in \mathbb{R}$$

of a quasilinear autoregressive model

$$y_t = \sum_{j=1}^{m} a_j g_j \{y_{t-k}\}_{k=1}^m + \sum_{j=1}^{n} b_j x_{tj} + \varepsilon_t, \quad t = 1, 2, \ldots, T \tag{1}$$

The work is supported by Act 211 Government of the Russian Federation, contract No. 02.A03.21.0011.

N. Olenev et al. (Eds.): OPTIMA 2020, CCIS 1340, pp. 15–26, 2020.
https://doi.org/10.1007/978-3-030-65739-0_2

by up-to-date information about of values of state variables (i.e. endogenous variables) $\{y_t \in \mathbb{R}\}_{t=-m+1}^{T}$ and values of control variables (i.e., exogenous variables) $\{x_{t1}, x_{t2}, \ldots, x_{tn} \in \mathbb{R}\}_{t=1}^{T}$ at time instants t, where $g_j : \mathbb{R}^m \to \mathbb{R}$, $j = 1, 2, \ldots m$ are given functions, and $\{\varepsilon_t \in \mathbb{R}\}_{t=1}^{T}$ are random errors.

Following the work [16], let us introduce the new notation

$$
A = \begin{bmatrix} a_1 \\ a_2 \\ \cdots \\ a_m \\ b_1 \\ b_2 \\ \cdots \\ b_n \end{bmatrix} ; \quad X_t = \begin{bmatrix} g_1\{y_{t-k}\}_{k=1}^{m} \\ g_2\{y_{t-k}\}_{k=1}^{m} \\ \cdots \\ g_m\{y_{t-k}\}_{k=1}^{m} \\ x_{1t} \\ x_{2t} \\ \cdots \\ x_{nt} \end{bmatrix} ; \quad t = 1, 2, \ldots, T; \quad N = n + m,
$$

in order to make mathematical expressions less cumbersome.

In these terms, Eq. (1) takes the form

$$
y_t = A^\mathsf{T} X_t + \varepsilon_t, \quad t = 1, 2, \ldots, T. \tag{2}
$$

As a rule, system (1), and system (2), are incompatible, and methods for optimizing the loss function of a suitable form are used to solve it. The most known method for determining the coefficients of the regression equation is the least squares method (LSM)

$$
A^* = \arg \min_{A \in \mathbb{R}^N} \sum_{t=1}^{T} \left(A^\mathsf{T} X_t - y_t\right)^2. \tag{3}
$$

LSM is the parametric method and requires a number of strict restrictions: the determinism of variables, the independence and normality of the distribution of measurement errors [3, 7, 12]. Even minor violations of these prerequisites critically reduce the effectiveness of LSM estimations [6].

If we allow errors in the measured values of endogenous variables y_t, $t = 1, 2, \ldots, T$, then their presences in the values of the functions $g_j\{y_{t-k}\}_{k=1}^{m}$ are obvious. Moreover, these errors have to be mutually correlated, and have probability distributions different from the normal distribution. This makes the classical solution schemes based on the LSM and its variations ineffective. The estimates of the autoregressive equation factors is substantially complicated by the ill-conditioning of the equation systems representing the necessary conditions for the minimum of the sum of the squares of the deviations, while the estimates become insolvent.

An alternative to LSM is the Least Deviations Method (LDM) [2, 4, 5, 12, 20]

$$
A^* = \arg \min_{A \in \mathbb{R}^N} \sum_{t=1}^{T} \left|A^\mathsf{T} X_t - y_t\right|. \tag{4}
$$

Its possible generalizations are Weighted Least Deviation Method (WLDM) [13]

$$A^* = \arg \min_{A \in \mathbb{R}^N} \sum_{t=1}^{T} \left(p_t \left| A^\mathsf{T} X_t - y_t \right| \right) \text{ for prefixed } p_t \in \mathbb{R}^+, \tag{5}$$

and the Generalised Least Deviations Method (GLDM)[19, 23]

$$A^* = \arg \min_{A \in \mathbb{R}^N} \sum_{t=1}^{T} \rho \left(\left| A^\mathsf{T} X_t - y_t \right| \right) \text{ for convex up differentiable function } \rho(*).$$
$$\tag{6}$$

Problems (4) and (5) are piecewise linear programming problems and may be reduced to a linear programming problem. Algorithms for the exact solution of LDM estimation problems (4) are described in [22]. This algorithm has computational complexity $O(N^2 T^2 + N^4 T \ln T + N^2 T \ln^2 T)$. But this algorithm application for solving WLDM problem (5) and selecting weight factors for it are not clear.

Problem (6), i.e. problem of GLDM estimation, is a concave optimization problem. GLDM estimates are robust to the presence of a correlation of values in $\{X_{jt} : t = 1, 2, \ldots, T; \ j = 1, 2, \ldots, N\}$, and (with appropriate settings) are the best for probability distributions of errors with heavier (than normal distribution) tails [19]. All the above shows the feasibility of solving the identification problem (1) by method (6).

The established in [14–16] results allow us to reduce the problem of determining GLDM estimation to an iterative procedure with WLDM estimates.

The method to increase effectiveness of the QRE identification algorithm is proposed in this paper. The proposed method is based on the solution of the modified dual linear programming problem, and has a computational complexity of not more than $O(N^3 \cdot T + N \cdot T^2)$ by taking into account the specifics of this problem.

In Sect. 2, the way of reducing WLDM problem (5) to simple structure problem (13) solved by algorithm **PrGrad** with computational complexity $O(T \cdot N^2 + T^2)$ is described. Algorithm **WLDM-estimator** to solve WLDM problem (5) with problem (13) solutionis is suggested in Sect. 3. Section 4 is devoted to the algorithm of GLDM estimation [14] in terms of this paper. The results of computational experiments for problem of identifying the recurrence equation of the stock market index in Iraq are presented in Sect. 5. All the results are summarized in Sect. 6.

2 WLDM Estimation Problem

WLDM estimation algorithm to identify Eq. (1) leads to the solution of the optimization problem (5) for given $X_t \in \mathbb{R}^N, \quad y_t, p_t \in \mathbb{R}, \quad t = 1, 2, \ldots T.$

This problem is equivalent to the linear programming problem

$$\sum_{t=1}^{T} p_t z_t \to \min_{A \in \mathbb{R}^N, \, z \in \mathbb{R}^T}, \tag{7}$$

$$A^\mathsf{T} X_t + z_t \geq y_t, \ t = 1, 2, \ldots, T, \tag{8}$$

$$-A^\mathsf{T} X_t + z_t \geq -y_t, \ t = 1, 2, \ldots, T, \tag{9}$$

Dual for (7)–(9) problem has form

$$\sum_{t=1}^{T} (u_t - v_t) y_t \to \max_{u, v \in \mathbb{R}^T}, \tag{10}$$

$$\sum_{t=1}^{T} X_{jt} (u_t - v_t) = 0, \ j = 1, 2, \ldots, N, \tag{11}$$

$$u_t + v_t = p_t, \quad u_t, v_t \geq 0, \quad i = 1, 2, \ldots, T, \tag{12}$$

Let us introduce variables $w_t = u_t - v_t$, $t = 1, 2, \ldots, T$. Conditions (12) imply

$$u_t = \frac{p_t + w_t}{2}, \quad v_t = \frac{p_t - w_t}{2}, \quad t = 1, 2, \ldots, T.$$

It is following (12) that $-p_t \leq w_t \leq p_t$, $t = 1, 2, \ldots, T$. Therefore optimal value of the problem (10)–(12) is equal to optimal value of problem

$$\sum_{t=1}^{T} w_t \cdot y_t \to \max_{w \in \mathbb{R}^T}, \tag{13}$$

$$\sum_{t=1}^{T} X_{jt} w_t = 0, \ j = 1, 2, \ldots, N, \tag{14}$$

$$-p_t \leq w_t \leq p_t, \ t = 1, 2, \ldots, T. \tag{15}$$

The admissible set of problem (13) is intersection of T-dimensional cuboid (15) and $(T - N)$-dimensional linear subspace (14). Let us consider the usage of the gradient projection method [11,21] to solve problem (13).

The algorithm is described below.

PrGrad

Input: $X = \{X_t \in \mathbb{R}^N\}_{t \in T}, \quad p \in \mathbb{R}^{+T}, \quad y \in \mathbb{R}^T$.
Output: $w^* = \arg\max_{w \in \mathbb{R}^T} \sum_{i=1}^{T} w_i \cdot y_i, \quad R = \{t \in T : |w_t^*| = p_t\}$.

Step 1. Initialization.
$\quad w := \{w_i = 0 : i = 1, 2, \ldots, T\}; \ R := \emptyset; \ /* \text{Starting point} */$
$\quad g := y - X^\mathsf{T} \left(\left([XX^\mathsf{T}]^{-1} X \right) y \right); \ /* \text{Projection of gradient} */$

Step 2. Current iteration
   Do
$$(\alpha_*, t_*) := \arg\max_{\alpha, t} \{\alpha > 0 : -p_t \leq w_t + \alpha g_t \leq p_t\} \text{ /*Step parameters*/}$$
$$w := w + \alpha_* g; \quad g_{t_*} := 0; \quad R := R \cup \{t_*\};/\text{* Next Point */}$$
   While $\quad (\alpha_* \neq 0)$ /* The stop criteria */
Return $w^* = w$, $R^* = R$.
End of PrGrad

Theorem 1. *The algorithm* **PrGrad** *solves the problem* (13)–(15). *Its compu-tational complexity does not exceed the quantity* $O(T \cdot N^2 + T^2)$.

Proof. Each k-th iteration of the algorithm consists of an admissible movement from the current point $w^{(k)}$ to the next point $w^{(k+1)} = w^{(k)} + \alpha_* g^{(k)}$ in the direc-tion $g^{(k)}$ equal to projection of objective function gradient y to intersection set of the equations system (14) solutions, and set of solutions for active constraints system

$$R^{(k)} = \left\{ t : \left| w_t^{(k)} \right| = p_t \right\}$$

from inequality system (15). Obviously, the points of the sequence constructed in this way are valid.

It is following from the description of the algorithm that

- Slater condition holds because $w = 0$ is the interior point of the problem admissible set,
- the necessary condition $(\alpha_* = 0)$ of the local maximum holds,
- the sufficient condition $w_t^{(k)} y_t > 0$ for all $t \in R^{(k)}$ (i.e. the gradient y cannot be represented as a non-negative linear combination of gradients of active constraints) holds,

take place after cycle **While** termination. Therefore, in accordance with the Kuhn-Tucker theorem, solution $w^{(k)}$ is optimal. This proves the first proposition of the theorem.

Let's estimate the computational complexity of **Step1**. The computational complexity of multiplying the $(N \times T)$-matrix by the $(T \times N)$-matrix, i.e. com-puting the matrix in square brackets does not exceed the quantity $O(T \cdot N^2)$. The complexity of the inverse of the obtained $(N \times N)$-matrix does not exceed the value $O(N^3)$. Consequently, computational complexity of calculation algorithm

$$g = y - X^\top \left(\left([XX^\top]^{-1} X \right) y \right)$$

does not exceed $O(T \cdot N^2)$ because $T > N$.

Computational complexity of cycle **While** body is no more $O(T)$, and it is executed no more than T times. Therefore computational complexity of **Step1** is no more than $O(T^2)$, and computational complexity of algorithm **PrGrad** is no more than $O(T \cdot N^2 + T^2)$. This proves the second proposition of the theorem.

3 Improved WLDM-Estimation Algorithm

If (w^*, R^*) result of algorithm **PrGrad** then w^* is optimal solution of problem (13)–(15), and optimal solution of problem (10)–(12) is equal

$$u_t^* = \frac{p_t + w_t^*}{2}, \quad v_t^* = \frac{p_t - w_t^*}{2}, \quad t = 1, 2, \ldots, T.$$

The complementarity condition for a pair of mutually dual problems (7)–(9) and (10)–(12) implies

$$y_t = \begin{cases} (A^*)^\mathsf{T} X_t, & \text{if } t \notin R^*, \\ (A^*)^\mathsf{T} X_t + z_t^*, & \text{if } (t \in R^*, w_t^* = p_t), \\ (A^*)^\mathsf{T} X_t - z_t^*, & \text{if } (t \in R^*, w_t^* = -p_t), \end{cases} \quad t = 1, 2, \ldots, T. \tag{16}$$

In fact, solution (A^*, z^*) of linear equation system (16) is the optimal dual solution of problem (13)–(15) and an optimal solution of the problem (7)–(9), so the following theorem is proved.

Theorem 2. *Let w^* be optimal solution of problem (13)–(15), let (A^*, z^*) be solution of linear equation system (16), then A^* is the optimal solution to the problem (5).*

Algorithm of WLDM estimation is described below.

WLDM-estimator

Input: $X = \{X_t \in \mathbb{R}^N\}_{t \in T}$, $p \in \mathbb{R}^{+T}$, $y \in \mathbb{R}^T$.
Output:
Estimation $A^* \in \mathbb{R}^N$ of autoregressive equation (5) factors;
Residuals $z \in \mathbb{R}^T$ of equations system (16).

Step 1. Let $(w^*, R^*) = \mathbf{PrGrad}\,(X, p, y)$.
Step 2. Let A^* be a solution of linear equations

$$(A^*)^\mathsf{T} X_t = y_t, \; t : \; t \notin R^*.$$

Step 3 Let $z^* = (A^*)^\mathsf{T} X - y$ be residuals of equations system (16).
Return (A^*, z^*).
End of WLDM-estimator

Theorem 3. *The algorithm **WLDM-estimator** solves the problem (5). Its computational complexity does not exceed the quantity $O(T \cdot N^2 + T^2)$.*

Proof. Let us consider the system of equations

$$(A^*)^\mathsf{T} X_t = y_t, \qquad t : \; t \notin R^* \tag{17}$$

that is solved at **Step 2**. Number of equations of this system is equal to the number $K = |\{t : \; t \notin R^*\}|$ of the unfixed dual variables by algorithm **PrGrad**.

It is obvious that the number M of coordinates with cuboid extreme values is not less than $(T - N)$ at the extreme points of intersection of the T-dimensional cuboid (15) and the $(T - N)$-dimensional linear manifold (14). Therefore, the number of free dual variables is equal to $K = T - M \leq N$.

From the description of the algorithm **PrGrad**, it can be seen that the set R^* has a minimum power, i.e. it is equal to $(T - N)$.

Existence of a solution to problem (5) and also equivalence of the problems (5) and (7)–(9) imply that system (17) is compatible.

Thus, system (17) is the compatible system of N equations with N unknowns. The solution of this system can be found by the Jordan-Gauss algorithm in a time not exceeding $O(N^3)$. It follows from Theorem 1 that computational complexity of **Step 1** does not exceed $O(T \cdot N^2 + T^2)$. The validity of the theorem by virtue of the inequality $T > N$.

4   GLDM Estimation Problem

Problem (6) of GLDM estimation is a concave optimization problem. GLDM-estimates are robust to the presence of a correlation of values in $\{X_{jt} : t = 1, 2, \ldots, T;\ j = 1, 2, \ldots, N\}$, and (with appropriate settings) like the best for probability distributions of errors with heavier (than normal distribution) tails [19]. The above shows the feasibility of solving the identification problem (1) by method (6).

The established in [14,15] results allow us to reduce the problem of determining GLDM estimation to an iterative procedure with WLDM estimates.

Let us consider the algorithm of GLDM estimation [14] in terms of this paper.

GLDM-estimator

Input: number of measures $T \in \mathbb{N}$; $(N \times T)$ matrix $X = \{X_t \in \mathbb{R}^N\}_{t \in T}$; convex upwards twice continuously differentiable function $\rho(*) : \mathbb{R}^+ \to \mathbb{R}^+$.
Output:
estimation of coefficients $A^* \in \mathbb{R}^N$ of autoregressive equations (1) and/or (2).

Step 1.
For all $t \in \{1, 2, \ldots, T\}$ **do** $p_t = 1$;
$k := 0$; $\left(A^{(k)}, z^{(k)}\right) := $ **WLDM-estimator** $(X,\ p,\ y)$.

Step 2.
Do
    For all $t \in \{1, 2, \ldots, T\}$ **do** $p_t := \rho'(z_t^{(k)})$;
    $k := k + 1$; $\left(A^{(k)}, z^{(k)}\right) := $ **WLDM-estimator** $(X,\ p,\ y)$.
While $\left(A^{(k)} \neq A^{(k-1)}\right)$.

Return $A^* := A^{(k)}$.
End of GLDM-estimator

The description of the algorithm **GLDM-estimator** shows that its computational complexity is proportional to the computational complexity of the algorithm for solving of WLDM problem (5).

Multiply computational experiments show that the average number of iterations of algorithm **GLDM-estimator** is equal to the number of coefficients in the identified equation. If this hypothesis is true then computational complexity in solving practical problems does not exceed $O(N^3 T + N T^2)$.

5   Computational Experiment

Computational experiments involving the construction of the solution for Cauchy problem to some quasi-linear difference equation and subsequent identification of this equation by the constructed solution shows the high quality of the proposed algorithm [10]. Here we present the results of computational experiments to solve the problem of identifying the unknown recurrence equation of the stock market index in Iraq According to the original data from the site "ISX-IQ.net" [1].

Let us consider three mathematical models:

– model **10-Freedom**

$$y_t = a^{(0)} + \left(a_1^{(1)} y_{t-1} + a_2^{(1)} y_{t-2} \right) + \left(a_{11}^{(2)} y_{t-1}^2 + a_{12}^{(2)} y_{t-1} y_{t-2} + a_{22}^{(2)} y_{t-2}^2 \right)$$
$$+ \left(a_{111}^{(3)} y_{t-1}^3 + a_{112}^{(3)} y_{t-1}^2 y_{t-2} + a_{122}^{(3)} y_{t-1} y_{t-2}^2 + a_{222}^{(3)} y_{t-2}^3 \right), \quad t = 2, 3, \ldots, T; \quad (18)$$

– model **5-Freedom**

$$y_t = \left(a_1^{(1)} y_{t-1} + a_2^{(1)} y_{t-2} \right) + \left(a_{11}^{(2)} y_{t-1}^2 + a_{12}^{(2)} y_{t-1} y_{t-2} + a_{22}^{(2)} y_{t-2}^2 \right),$$
$$t = 2, 3, \ldots, T; \quad (19)$$

– model **2-Freedom**

$$y_t = \left(a_1^{(1)} y_{t-1} + a_2^{(1)} y_{t-2} \right), \quad t = 2, 3, \ldots, T. \quad (20)$$

Input data for algorithm **GLDM-estimator** are the following.

– As endogenous variables are used of one hundred and eighty one (181) consecutive counts of Iraq stock market daily index $Y_t : \ t = 0, 1, 2, \ldots, 180$ presented at Fig. 1.
– Set of exogenous variables is empty.
– Convex upwards twice continuously differentiable function

$$\rho(*) : \ \mathbb{R}^+ \to \mathbb{R}^+ : \ \rho(z) = \arctan z.$$

Identification results are presented in Table 1. It shows that

– model **2-Freedom** gives the lowest value of the loss function;
– model **5-Freedom** gives the maximum value of the loss function;
– coefficients of all models are significant;

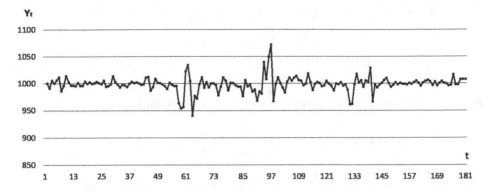

Fig. 1. Observed time series

Table 1. The identification results

Factors	Model		
	10-Freedom	5-Freedom	2-Freedom
$a^{(0)}$	0.000000e+000		
$a_1^{(1)}$	5.026171e+001	−6.008402e+000	3.134622e−001
$a_2^{(1)}$	4.501042e+001	8.017241e+000	6.856777e−001
$a_{11}^{(2)}$	1.676391e−001	3.732025e−003	
$a_{12}^{(2)}$	−2.386618e−001	−1.651187e−003	
$a_{22}^{(2)}$	8.436132e−002	−3.090063e−003	
$a_{111}^{(3)}$	−7.228885e−005		
$a_{112}^{(3)}$	5.538078e−005		
$a_{122}^{(3)}$	5.964459e−005		
$a_{222}^{(3)}$	−4.982450e−005		
Value of loss function	3.914645e+001	4.118230e+001	3.624738e+001
GLDM number of iterations	10	5	6

– the number of iterations of the algorithm is approximately proportional to the number N of model coefficients.

Most likely, the increasing of the loss function value with an increasing of freedom degrees number N is due to a cubic $(O(N^3))$ increasing our calculations, and therefore significant accumulated calculation errors but not model quality.

The purpose of identifying the equations (18)–(20) is enabling to use the model values of endogenous variables for predicting possible values of corresponding endogenous variables in the future.

Let K be the prediction horizon, and $y_t = F(y_{t-1}, y_{t-2})$ be one of the model functions (18)–(20). The modal values of endogenous variables $\tilde{y}_t$ for time counts $t, t+1, \ldots, t+K$ are

$$\tilde{y}_t^{(K)} = F\left(y_{t-1}, y_{t-2}\right),$$

$$\tilde{y}_{t+1}^{(K)} = F\left(\tilde{y}_t^{(K)}, y_{t-1}\right),$$

$$\tilde{y}_{t+k}^{(K)} = F\left(\tilde{y}_{t+k-1}^{(K)}, \tilde{y}_{t+k-2}^{(K)}\right), \quad k = 2, 3, \ldots, K.$$

Average prediction error for prediction horizon K is equal to

$$E(K) = \frac{1}{T-K+1} \cdot \sum_{t=0}^{T-K} \left(\tilde{y}_{t+K}^{(K)} - y_{t+K}\right),$$

and average absolute prediction error for prediction horizon K is equal to

$$D(K) = \frac{1}{T-K+1} \cdot \sum_{t=0}^{T-K} \left|\tilde{y}_{t+K}^{(K)} - y_{t+K}\right|.$$

Table 2. The average error over the forecast horizon

Forecast horizon K	Model					
	10-Freedom		5-Freedom		2-Freedom	
	$E(K)$	$D(K)$	$E(K)$	$D(K)$	$E(K)$	$D(K)$
1	4.296E−02	6.841E+00	9.050E−01	7.565E+00	−8.621E−01	3.323E+00
2	9.617E−01	8.567E+00	1.660E+00	8.953E+00	−1.531E+00	9.773E+00
4	5.142E−01	8.256E+00	7.483E−01	8.359E+00	−2.820E+00	1.282E+01
8	4.852E−01	8.282E+00	8.851E−01	8.281E+00	−5.446E+00	1.318E+01
16	5.350E−01	8.459E+00	9.354E−01	8.453E+00	−1.063E+01	1.700E+01
32	7.543E−01	8.982E+00	1.154E+00	8.986E+00	−2.076E+01	2.434E+01
64	−2.572E−01	8.278E+00	1.413E−01	8.268E+00	−4.131E+01	4.170E+01
128	−1.384E−01	6.264E+00	2.517E−01	6.224E+00	−7.695E+01	7.695E+01

Table 2 shows values $E(K)$ and $D(K)$ for the constructed models. Prediction results are presented in Table 2 show that

- for the linear model 2-Freedom the dependence of errors $E(K)$ and $D(K)$ on the value K of the forecast horizon is monotonous;
- for nonlinear models 5-Freedom and 10-Freedom the dependence of errors $E(K)$ and $D(K)$ on the value K of the forecast horizon is non-monotonic, in particular, the prediction errors for the extreme values of $K \in \{1, 128\}$ less than for intermediate values $K \in \{2, 4, 8, 16, 32, 64\}$;
- consistency of the 5-Freedom and 10-Freedom models for all K;
- 10-Freedom model gives the smallest prediction errors.

The computational experiment conducted in this paper is based on the usage of standard 32-bit numeric data types. Perhaps increasing the order of the recurrence equation and the accuracy of the calculations [17,18] will give better results.

6   Conclusion

Algorithm **WLDM-estimator** to solve the WLDM-estimation problem is proposed. Computational complexity of the algorithm does not exceed the quantity $O(N^2T + T^2)$, where N is the number of coefficients in the studied equation, T is the number of observed readings. It is possible to use the QRE identification algorithm **GLDM-estimator** based on generalized least deviations method. **GLDM-estimator** solves the sequence of the WLDM-estimation problems. The computational complexity of solving practical problems does not exceed $O(N^3T + NT^2)$.

Results of computational experiments to solve the problem of identifying the recurrence equation of the stock market index in Iraq by original data from the site "ISX-IQ.net" show the possibility to apply a second order quasilinear recurrence equation with quadratic nonlinearity for these purposes.

To solve more complex problems, for example such as presented in papers [8,9], increasing the order of the recurrence equation and the accuracy of the calculations [17,18] are required.

References

1. www.isx-iq.net. (in Arabic)
2. Akimov, P.A., Maslov, A.I.: Levels of nonoptimality of the Wciszfcld algorithm in the least modules method. Autom. Remote Control **71**(2), 172–184 (2010). https://doi.org/10.1134/S0005117910020025
3. Ayvazyan, S.A., Enyukov, I.S., Meshalkin, L.: Prikladnaya statistika: Issledovanie zavisimostey. Appl. Stat.: Dependency Stud. M: Finansy i statistika (1985). (in Russian)
4. Bloomheld, P., Steiger, W.L.: Least Absolute Deviations: Theory, Applications, and Algorithms. Birkhauser, Boston-Basel-Stuttgart (1983)
5. Dielman, E.T.: Least absolute value regression: recent contributions. J. Stat. Comput. Simul. **75**(4), 263–286 (2003). https://doi.org/10.1080/0094965042000223680
6. Huber, P., Ronchetti, E.: Robust Statistics, 2nd edn. Wiley, Hoboken (2009)
7. Mandelbrot, B.B.: New methods in statistical economics. J. Polit. Econ. **71**, 421–440 (1963)
8. Mandelbrot, B.B.: The variation of certain speculative prices. In: The Random Character of Stock Market Prices, p. 510. M.I.T. Press, Cambridge (1964)
9. Mandelbrot, B.B.: The Fractal Geometry of Nature. W.H. Freeman, New York (1982)
10. Mezaal, Y.A.: Chislennoye issledovaniye determinirovannoy modeli kvazilineynogo analiza vremennykh ryadov [numerical study of the deterministic model of quasilinear analysis for time series]. Nauchnyy obozrevatel [Sci. Obser.] (9), 5–11 (2019). (in Russian). http://nauchoboz.ru/wp-content/uploads/2019/10/Nauchoboz-9-2019.pdf
11. Minoux, M.: Programmation mathematique: theorie et algorithmes [Mathematical programming: theory and algorithms]. Bordas et. C.N.P.T. - E.N.S.T., Paris (1989). (in French)

12. Mudrov, V.I., Kushko, V.L.: Melody obrabotki izmereniy: kvazipravdopodob-nye otsenk (Measurement Processing Methods: Quasi-Truth Estimates), 3rd edn. Levand, Moscow (2014). (in Russian)
13. Pan, J., Wang, H., Qiwei, Y.: Weighted least absolute deviations estimation for arma models with infinite variance. Economet. Theory **23**(3), 852–879 (2007)
14. Panyukov, A.V., Mezaal, Y.A.: Stable estimation of autoregressive model parameters with exogenous variables on the basis of the generalized least absolute deviation method. In: IFAC-PapersOnLine, vol. 51, pp. 1666–1669 (2018). https://doi.org/10.1016/j.ifacol.2018.08.217. Open access
15. Panyukov, A.V., Mezaal, Y.A.: Parametricheskaya identifikatsiya kvazilineynogo raznostnogo uravneniya [parametric identification of the quasilinear difference equation]. Bull. South Ural State Univ. Ser. Math. Modell. Program. Comput. Softw. **11**(4), 32–38 (2019). https://doi.org/10.14529/mmp180104. (in Russian)
16. Panyukov, A.V., Mezaal, Y.A.: Parametricheskaya identifikatsiya kvazilineynogo raznostnogo uravneniya [parametric identification of the quasilinear difference equation]. Bull. South Ural State Univ. Ser. Math. Mech. Phys. **11**(4), 32–38 (2019). https://doi.org/10.14529/mmph190404. (in Russian)
17. Panyukov, A.: Scalability of algorithms for arithmetic operations in radix notation. Reliable Comput. **19**, 417–434 (2015). http://interval.louisiana.edu/reliable-computing-journal/volume-19/reliable-computing-19-pp-417-434.pdf
18. Panyukov, A., Gorbik, V.: Using massively parallel computations for absolutely precise solution of the linear programming problems. Autom. Remote Control **73**(2), 276–290 (2012). https://doi.org/10.1134/S0005117912020063
19. Panyukov, A., Tyrsin, A.: Stable parametric identification of vibratory diagnostics objects. J. Vibroeng. **10**(2), 142–146 (2008). http://elibrary.ru/item.asp?id=14876532
20. Powell, J.L.: Least absolute deviations estimation for the censored regression model. J. Econometr. **25**, 303–325 (1984)
21. Rosen, J.B.: The gradient projection method for nonlinear programming, part 1: linear constraints. J. Soc. Ind. Appl. Math. **8**, 181–217 (1960)
22. Tyrsin, A.N., Azaryan, A.F.: Exact evaluation of linear regression models by the least absolute deviations method based on the descent through the nodal straight lines. Bull. South Ural State Univ. Ser. "Math. Mech. Phys." **10**(2), 47–56 (2018). https://doi.org/10.14529/mmph180205. (in Russian)
23. Tyrsin, A.: Robust construction of regression models based on the generalized least absolute deviations method. J. Math. Sci. **139**(3), 6634–6642 (2006). https://doi.org/10.1007/s10958-006-0380-7

Optimization Algorithm
for Approximating the Solutions Set
of Nonlinear Inequalities Systems
in the Problem of Determining the Robot
Workspace

Larisa Rybak[ID], Dmitry Malyshev$^{(\boxtimes)}$[ID], and Elena Gaponenko[ID]

Belgorod State Technological University named after V.G. Shukhov, Belgorod, Russia
rlbgtu@gmail.com

Abstract. This paper is devoted to the problem of determining the workspace of robots. We consider an approach to the development of a numerical method for approximating the set of solutions of a system of nonlinear inequalities based on the concept of non-uniform coverings. An approach is proposed based on the transformation of non-uniform covering sets into a set of partially ordered sets of integers to reduce computational complexity. An algorithm for transforming boxes of a covering set is presented. The approach has been tested for a 3-RPS robot. The results of the mathematical simulation and analysis of the effectiveness of the proposed approach based on an estimate of the reduction in the amount of numbers describing the covering set are presented.

Keywords: Robot workspace · Parallel robot · Non-uniform covering · Optimization algorithm

1 Introduction

Deterministic methods allow us to solve global optimization problems with an estimate of the value of the approximate solution found from the optimum. However, the actual problem in applying these methods is often considerable computational complexity. The development of approaches to reduce it is an urgent task.

The method of non-uniform coverings [1] is one of famous deterministic method. It was proposed by Yu.G. Evtushenko in 1971 to solve problems with box constraints. This method can be easily automated and applied to solve a number of various problems, including in the field of robotics. One of them is the definition of the workspace of robots, within which there should be a working tool when performing technological operations. The workspace is one of the

This work was supported by the Russian Science Foundation, the agreement number 16-19-00148.

key characteristics of robots, including parallel ones. The issues of structural synthesis, methods for studying the workspace, and optimizing the trajectory of movement of such mechanisms are considered in detail in [2–4]. The application of the method of non-uniform coverings to determine the workspace is considered in [5–8]. In [7], a comparison of two approaches is considered, one of which is based on the use of a system of inequalities to describe the design constraints of the robot, and the other on the use of a system of equations. Using the system of equations $g_j(x) = 0, j \in \{1, m\}$, the workspace is described by the set Q_E of n-dimensional boxes P_i, including the set of solutions of the system, i.e., $Q_E = \bigcup_{i \in I} P_i$. x is n-dimensional vector of variables. For each P_i from Q_E, the system of equations holds:

$$\begin{cases} \max_{j=1,\ldots,m} \min_{x \in P_i} g_j(x) \leq 0, \\ \min_{j=1,\ldots,m} \max_{x \in P_i} g_j(x) \geq 0, \\ d(P_i) \leq \delta. \end{cases} \tag{1}$$

Using the system of inequalities $g_i(x) \leq 0, j \in \{1, m\}$ the workspace is described by the union of two sets: $Q_E = Q_I \cup Q_J$, where Q_I is the inner approximation set that is included in the set of solutions of the system of inequalities, Q_J is the boundary set. For each P_i from Q_I, the following condition holds

$$\max_{j=1,\ldots,m} \max_{x \in P_i} g_j(x) \leq 0. \tag{2}$$

For each P_i from Q_J, the system of inequalities holds:

$$\begin{cases} \max_{j=1,\ldots,m} \max_{x \in P_i} g_j(x) > 0, \\ \max_{j=1,\ldots,m} \min_{x \in P_i} g_j(x) \leq 0, \\ d(P_i) \leq \delta. \end{cases} \tag{3}$$

One of the tools for implementing the method is iterative bisection. With each division, the box decreases by 2 times, respectively, the ratio of the sizes of the initial box and one of the boxes forming an approximation of the workspace can be estimated by the degree of division d, and the ratio itself is 2^d. With an increase in the degree of division d, the number of boxes, the combination of which describes the workspace, increases. Due to the increase in computational complexity for processing an approximated workspace of higher accuracy, the problem arises of reducing it by transforming the resulting set of boxes describing the workspace. As part of this work, an approach is proposed for transforming a covering set obtained using the method of non-uniform coverings into a partially ordered set of integers. It includes two components: reducing the number of boxes and the transition from the space of real numbers to the space of integers. Application of the proposed approach and assessment of its effectiveness is considered on the problem of determining the workspaces of the 3-RPS robot.

2   Covering Sets Transformation to Partially Ordered Integer Sets

Consider the set of boxes that form the workspace. We introduce the following notation: δQ_E is the boundary of the set Q_E, $Q_A = \delta Q_E \cup (R^n / Q_E)$ is the outer region for which the condition holds

$$\max_{j=1,\ldots,m} \min_{x \in P_i} g_j(x) \leq 0. \tag{4}$$

Proposition 1. For any point $x = (x_1, .., x_n)$, $x \in Q_E$, there exist points $a = (a_1, .., a_n)$, $a \in Q_J$ and $b = (b_1, .., b_n)$, $b \in Q_J$, for which the following system is satisfied:

$$\begin{cases} a_1 \leq x_1 \leq b_1, \\ x_i = a_i = b_i, i \in \{2, n\}. \end{cases} \tag{5}$$

Proof. Let us prove the statement "by contradiction". Assume that in Q_J there is no point a for which system (5) is satisfied. In this case, given that Q_E is a finite set, there is a point a, $a \in \delta Q_E$, with $c \notin Q_J$ for which the system is satisfied. Since $Q_E = Q_I \cup Q_J$, then the point $a \in Q_I$. Therefore, the condition $\max_{j=1,\ldots,m} g_j(x) \leq 0$ is fulfilled for it. Since $a \in \delta Q_E$, the condition $\max_{j=1,\ldots,m} g_j(x) > 0$ must be satisfied. Similarly, the conditions must be satisfied for point b. This contradicts the assumption and the statement is proved.

In other words, statement 1 shows that between the elements of the sets Q_I and Q_A the elements of the set Q_J are necessarily located. Therefore, it is possible to describe the workspace as a set of boxes using the following approach. We introduce the following notation for the boundaries of the box:

$$\underline{x_i} \leq x_1 \leq \overline{x_i}, i \in \{1, n\}. \tag{6}$$

We denote two subsets in the set Q_J as Q_{J1} and Q_{J2}. The subset Q_{J1} includes only those boxes at the boundary of which there is a point $x = (x_1, .., x_n)$ for which the system is satisfied:

$$\begin{cases} x \in \delta Q_E, \\ x_1 = \underline{x_1}, \\ \underline{x_i} \leq x_1 \leq \overline{x_i}, i \in \{2, n\}. \end{cases} \tag{7}$$

The condition for the subset Q_{J2} is similar, but for it $c_1 = \overline{x_1}$.

It should be noted that the number m of boxes in the subsets Q_{J1} and Q_{J2} is equal to each box $P_{(k,J1)}$, $k \in 1$, m from Q_{J1} corresponds to the box $P(k, J_2)$, $k \in \{1, m\}$ from Q_{J2} with equal values $\underline{(x_i)}$, , $i \in \{2, n\}$ and $\overline{(x_i)}, i \in \{2, n\}$, while for all points $x = (x_1, .., x_n)$ for which the condition $x_{(i,J1)} < x_i < x_{(i,J2)}, i \in \{1, n\}$, $x \in Q_E$ is true. The set of Q_F is the set of boxes,

$$P_{(k,F)} = [x : \underline{x_{i,j_1,k}} \leq \underline{x_i} \leq \overline{x_{i,j_2,k}}, i], k \in \{1, m\}. \tag{8}$$

The n-dimensional box is described by 2n real numbers. The proposed approach to the transformation of boxes allows us to describe boxes with a smaller amount of numbers, while integers. First, we consider this concept in the general case of transforming the set of real numbers Y into the set of integers Z with the approximation accuracy δ (Fig. 1). For each of the points of the set, its coordinates in the space of integers are calculated:

$$x_i^{(j)'} = \left[\frac{x_i^j}{\delta}\right], x^{(j)} \in R, x^{(j)'} \in Z. \tag{9}$$

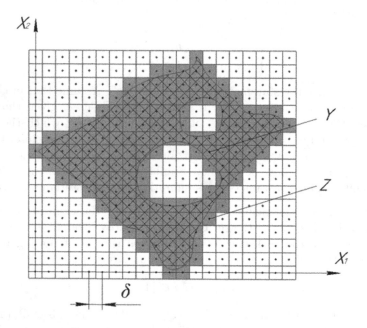

Fig. 1. The transformation of the cover set into the space of integers.

Moreover, the Hausdorff distance between the sets depends on δ

$$0 < h(Y, Z) \le \sqrt{\sum_i |\delta_i|^2}. \tag{10}$$

The integers of one of the coordinates are likewise combined into intervals, for each of which the values of the remaining coordinates are equal.

Let us consider the application of this concept to the transformation of the set of boxes Q_F to Q_Z. For Q_F and Q_Z the following holds:

$$h_{\max}(Q_F, Q_Z) = \sqrt{\sum |\delta_i|^2}. \tag{11}$$

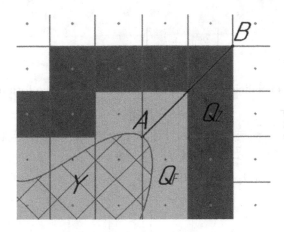

Fig. 2. Hausdorff distance between the sets Y and Q_Z.

The accuracy of determining the workspace allows us to estimate the Hausdorff distance between the sets Q_Z and Y (Fig. 2).

The maximum distance $h_{max}(Y, Q_Z)$ is defined as

$$h_{\max}(Y, Q_Z) = h_{\max}(Y, Q_F) + h_{\max}(Q_F, Q_Z). \tag{12}$$

Define $h_{max}(Y, Q_F)$. We denote the limitations of the original box when defining the workspace as

$$\underline{x_i^{(0)}} \leq x_i \leq \overline{x_i^{(0)}}, i \in \{1, n\}. \tag{13}$$

The sizes of the boxes of the set Q_J, taking into account (13), are defined as

$$\Delta_i = \frac{\overline{x_i^{(0)}} - \underline{x_i^{(0)}}}{2^d}. \tag{14}$$

Given that $h_{\max}(Y, Q_F) = h_{m}ax(Y, Q_j)$, we obtain:

$$h_{\max}(Y, Q_F) = \sqrt{\sum \left| \frac{\overline{x_i^{(0)}} - x_i^0}{2^d} \right|}. \tag{15}$$

We substitute (11) and (15) into (12):

$$h_{max}(Y, Q_Z) = \sqrt{\sum \left| \frac{\overline{x_i^{(0)}} - x_i^0}{2^d} \right|} + \sqrt{\sum |\delta_i|^2}. \tag{16}$$

In order to reduce the Hausdorff distance, we modify the formula

$$x_i' = \left\lceil \frac{2^d x_i + k_i}{\overline{x_i^{(0)}} - x_i^{(0)}} \right\rceil, x_i \in R, x_i' \in Z. \tag{17}$$

where k_i- bias coefficients, which are determined by the formula:

$$k_i = \left(\left\lceil \frac{2^d x_i^k}{\overline{x_i^{(0)}} - x_i^{(0)}} + 0,5 \right\rceil \right) \left(\frac{\overline{x_i^{(0)}} - x_i^{(0)}}{2^d} \right) - x_i^{(k)}, i \in \{1, n\}. \tag{18}$$

In this case, $h_{\max}(Q_F, Q_Z) = 0$ (Fig. 3).

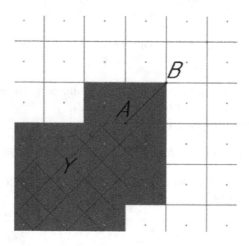

Fig. 3. The Hausdorff distance between the sets Y and Q_Z taking into account the displacement coefficients.

We introduce the following variables:

$$\underline{a_1} = \left\lceil \frac{2^d x_i + k_i}{\overline{x_i^{(0)}} - x_i^{(0)}} \right\rceil, \overline{a_i} = \left\lceil \frac{2^d \overline{x_i} + k_i}{\overline{x_i^{(0)}} - x_i^{(0)}} - 0,5 \right\rceil. \tag{19}$$

It is worth noting that an additional coefficient of 0.5 is added to exclude rounding of the upper boundary value to the next integer.

3   Application of the Developed Approach to Determine the Workspace of the 3-RPS Robot

Consider the application of the method of non-uniform coverings to determine the workspace of the planar 3-RPS mechanism (Fig. 4), which consists of three kinematic chains containing variable-length rods pivotally attached to a fixed base at the vertices of an equilateral triangle. The other ends of the rods are pivotally mounted at the vertices of an equilateral triangle on a movable platform. The abbreviation 3-RPS means that three chains is composed of a revolute joint, an actuated prismatic joint and a spherical joint mounted in series. The input coordinates are the rod lengths (l_1, l_2, l_3), the output coordinates are the position of the geometric center of the moving platform in Cartesian coordinates (x, y) associated with the center of the base of the mechanism, and its rotation angle (φ) relative to the axis perpendicular to the plane of the base. R and r are the radii of circles describing triangles and, respectively. This mechanism can be used to position the workpiece during machining.

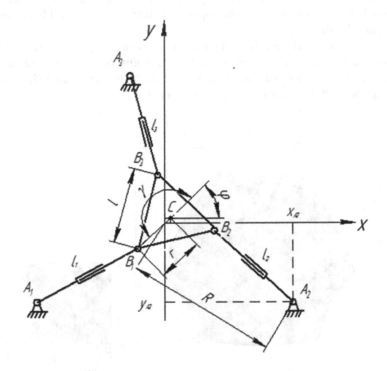

Fig. 4. Scheme of a planar 3-PRS mechanism.

Define the workspace of the 3-RPS mechanism. To do this, we introduce restrictions on the geometric parameters of the mechanism

$$l_{\min} \leq l_i \leq l_{\max}. \tag{20}$$

where $l_{\min}, l_{\max}$ are determined by the design parameters of the mechanism, l_i is the current length of the i-th rod, If the points Ai and Bi are located at the vertices of equilateral triangles, then the change in the length of the rods is determined by the formulas

$$l_1^2 = \left(x + \frac{r}{2}(s_\varphi - \sqrt{3}c_\varphi) + \frac{\sqrt{3}R}{2}\right)^2 + \left(y - \frac{r}{2}(\sqrt{3}s_\varphi + c_\varphi) + \frac{R}{2}\right)^2, \quad (21)$$

$$l_2^2 = \left(x + \frac{r}{2}(s_\varphi + \sqrt{3}c_\varphi) - \frac{\sqrt{3}R}{2}\right)^2 + \left(y + \frac{r}{2}(\sqrt{3}s_\varphi - c_\varphi) + \frac{R}{2}\right)^2, \quad (22)$$

$$i_3^2 = (x - rs_\varphi)^2 + (y + rc_\varphi - R)^2, \quad (23)$$

where $s_\varphi = \sin\varphi, c_\varphi = \cos\varphi$.

Algorithms for approximating the set of solutions of nonlinear inequalities were considered earlier in the authors' work [7]. To speed up the calculations, multithreaded calculations using the OpenMP library are used. This is considered in more detail in [9].

The simulation results for R = 400 mm, r = 50 mm, $l_{1,2,3}$ [200 mm, 500 mm] are presented in Fig. 5. The calculation time for approximation accuracy $\delta = 4$ mm, the grid dimension for calculating $16 \times 16 \times 16$ functions using parallel computing into 8 flows on a personal computer was 57 s.

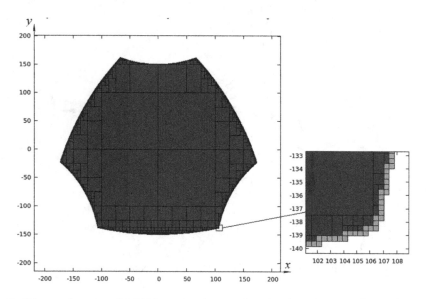

Fig. 5. The workspace of 3-PRS mechanism with a fixed angle $\varphi = 0°$: the blue area is the internal approximation, the yellow is the boundary

Dependence of the number of boxes on the degree of division for two-dimensional space (at a fixed angle $\varphi = 0°$) and three-dimensional space is shown in Table 1.

The table shows that, starting from the degree of division d = 6, the increase in the number of boxes approaches 2 for n = 2 and 4 for n = 3. We use the proposed approach to transforming the covering set.

Table 1. Dependence of the number of boxes on the degree of division.

Divide level d	n = 2		n = 3	
	Number of boxes	Increase in number	Number of boxes	Increase in number
0	1	–	1	0
1	4	4	8	8
2	4	1	16	2
3	4	1	32	2
4	4	1	64	2
5	16	4	334	5,094
6	34	2,125	1518	4,54
7	80	2,353	6410	4,246
8	168	2,1	26766	4,174
9	362	2,155	109390	4,089
10	732	2,022	441814	4,042
11	1504	2,055	1776510	4,021
12	3010	2,001	7124464	4,011
13	6050	2,010	28000055	4,005

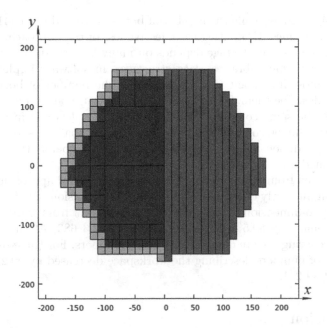

Fig. 6. The workspace before and after transformation of the covering set.

In Fig. 6 shows a visualization of the workspace symmetrical about the Y axis. The left half is described by the sets Q_I and Q_J, the right half is described by the set Q_F.

The transformation time for an approximation accuracy of $\delta = 4$ mm without using parallel computing was 3 min 30 s, that is, the total simulation time

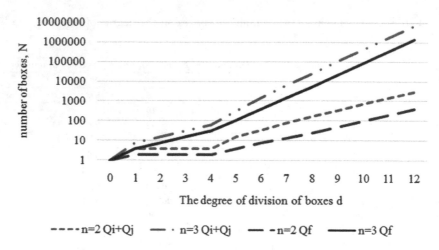

Fig. 7. Dependence of the number of boxes on the degree of division.

increased by 4.68 times, while the number of boxes decreased from 441 814 to 86 027, that is, 5.13 times. Evaluating the effectiveness of the approach, it is worth noting that the time of each stage depends on many factors, such as parallelization, the use of various libraries, data structures in software implementation. However, the most important thing is to reduce the number of boxes and the numbers that describe them. This allows to store and use better quality approximations with the same computational resources. The decrease in the number of boxes for two-dimensional and three - dimensional space is shown in Fig. 7. Number n means dimension, $\{Q_i+Q_j\}$ are 2 sets of boxes before transformation, Q_f - transformed set.

As can be seen from the figure, when using the proposed approach, the number of boxes significantly decreased. With a degree of division of 2^{12}, the number of boxes for two-dimensional space decreased by 86.7% from 3010 to 400, for three-dimensional - by 80.5% from 6 991 738 to 1 363 683. At the same time, boxes of the covering set are described by fewer numbers. For three-dimensional space, the set of numbers describing the workspace decreased by 90.2% from 41 950 428 to 4 099 241.

4 Conclusion

The proposed approach to the transformation of sets to reduce the number of covering boxes describing the workspace has shown its effectiveness. The approach has been tested on various dimensions of the problem. The approach's application for the degree of division of boxes 2^{12} allowed reducing the number of boxes of 3-RPS robot workspace from 6 991 738 to 1 363 683, that is, by 80.49%. The amount of numbers describing each of the transformed boxes has also been reduced. Thus, the overall reduction of numbers describing the workspace has been reduced from 41 950 428 to 4 099 241, that is, by 90.2%. A further area of research is to accelerate the transformation time.

References

1. Evtushenko, Y.G.: Numerical methods for finding global extrema (case of a non-uniform mesh). USSR Comput. Math. Math. Phys. **11**(6), 38–54 (1971)
2. Kong, X., Gosselin, C.M.: Type Synthesis of Parallel Mechanisms, vol. 33. Springer, Heidelberg (2007). https://doi.org/10.1007/978-3-540-71990-8
3. Merlet, J.P.: Parallel Robots, 2nd edn., pp. 269–285. Springer, Heidelberg (2007). https://doi.org/10.1007/1-4020-4133-0
4. Aleshin, A.K., Glazunov, V.A., Rashoyan, G.V., Shai, O.: Analysis of kinematic screws that determine the topology of singular zones of parallel-structure robots. J. Mach. Manuf. Reliab. **45**(4), 291–296 (2016). https://doi.org/10.3103/S1052618816040026
5. Posypkin, M.: Automated robot's workspace approximation. In: Journal of Physics: Conference Series, vol. 1163, no. 1, p. 012050. IOP Publishing (2019)
6. Evtushenko, Y., Posypkin, M., Rybak, L., Turkin, A.: Approximating a solution set of nonlinear inequalities. J. Global Optim. **71**(1), 129–145 (2017). https://doi.org/10.1007/s10898-017-0576-z
7. Malyshev, D., Posypkin, M., Rybak, L., Usov, A.: Approaches to the determination of the working area of parallel robots and the analysis of their geometric characteristics. Eng. Trans. **67**(3), 333–345 (2019)
8. Rybak, L.A., Behera, L., Malyshev, D.I., Virabyan, L.G.: Approximation of the workspace of parallel and serial structure manipulators as part of the multi-robot system. Bull. BSTU Named After V.G. Shukhov **8**, 121–128 (2019)
9. Malyshev, D.I., Posypkin, M.A., Gorchakov, A.Y., Ignatov, A.D.: Parallel algorithm for approximating the work space of a robot. Int. J. Open Inf. Technol. **7**(1), 1–7 (2019)

Parallel Global Optimization Algorithm with Uniform Convergence for Solving a Set of Constrained Global Optimization Problems

Vladislav Sovrasov[✉] and Konstantin Barkalov

Lobachevsky State University of Nizhny Novgorod, Nizhny Novgorod, Russia
{sovrasov.vladislav,konstantin.barkalov}@itmm.unn.ru

Abstract. This paper proposes an extension of global optimization algorithm for solving a set of problems with non-convex constraints. The uniform convergence inherent to the algorithm allows for the optimal distribution of the computational resources, since the number of errors in numerical solutions decreases at approximately equal rate for all problems processed by the algorithm. The algorithm assigns a priority level to each problem and, with every iteration, performs the computations of the objective functions for several problems in parallel. After the algorithm is terminated at any given time, it obtains solutions of similar quality for all the problems of the set. Such sets appear when the global optimization problem has a discrete parameter or, for example, when a multicriteria optimization problem is solved by scalarization techniques. The algorithm employs Peano-type space-filling curves for the reduction of the multidimensional optimization problems to the one-dimensional ones. The efficiency of the implemented algorithm was tested using the sets of artificially generated global optimization problems, as well as a series of problems obtained as a rescult of scalarization of a multicriteria optimization problem. Additional numerical experiments also confirmed the uniform convergence of the proposed method.

Keywords: Global optimization · Non-convex constraints · Uniform convergence · Derivative-free optimization · Parallel computing

1 Introduction

Global optimization problems with non-convex constraints are considered as one of the most difficult optimization problems. Finding of the global minimum of a function with several variables often appears to be more difficult than a local optimization in a thousand-dimensional space. The simplest gradient descent method could be sufficient for the latter, however, to *guarantee* that the global optimum is found, the optimization methods have to accumulate the information

This study was supported by the Russian Science Foundation, project No. 16-11-10150.

on the behavior of the objective function in the entire search domain [7,11,13,20]. In some cases, for instance, for DC or QP problems, more effective methods and even analytical optimality conditions could be applied [5,14], but these methods are barely applicable for black box functions (i.e., an analytical representation of the functions is not available).

Solving a series of such problems with limited computational resources is even more difficult: besides the search of the global extremum, the computational resources have to be distributed in such a way that the location of the global extremum would be estimated with approximately equal quality for all problems being solved at the same time. Usually, a series of q problems is solved either sequentially or in parallel by sets of $p \ll q$ problems where p is the number of parallel computational devices. Such an approach creates a situation when at every given moment there exist problems for which the global optimum is not obtained yet, while the optimum for the problems from the beginning of the set may be estimated even with excess precision.

This paper considers a parallel global optimization method developed at the Lobachevsky State University of Nizhny Novgorod for simultaneous solving of a set of problems [3] and in particular, its generalization to cover the non-convex constraints. The index scheme [16,20] which is applied to account for the constraints, allows operating with partially defined objective functions and constraints, and its efficiency is on par with similar approaches [1]. The convergence of the algorithm to the global optimizers in all problems was proved theoretically, the efficiency of the implemented algorithm was demonstrated on sets of problems generated by a special mechanism. This mechanism generates sets of problems with predefined dimensionality and predefined number of non-convex constraints [9]. Besides the artificially generated problems, the method was also tested on sets of multicriteria optimization problems with nonlinear constraints solved by criteria convolution method [6].

The paper is structured as follows. Section 2 provides the statement of the problem. Section 3 gives a description of the parallel optimization method. Subsection 3.1 describes the sufficient conditions of convergence of the considered method. The results of the numerical experiments confirming the efficiency of considered method are presented in Section 4. Conclusion gives a short summary of the results of the study and proposes the directions for further improvement of the software implementation of the considered method.

2   Statement of the Global Optimization Problem

This paper considers the following problem: to find the global minimum of an N-dimensional function $\varphi(y)$ in a hyperinterval D

$$\varphi(y^*) = \min\{\varphi(y) : y \in D\}, \tag{1}$$

$$D = \{y \in \mathbf{R}^N : a_i \leqslant x_i \leqslant b_i, 1 \leqslant i \leqslant N\}.$$

In order to produce an estimate of the global minimum based on a finite number of computations of the function values, the rate of variation of $\varphi(y)$ in D has to be limited. As a rule, the Lipschitz condition

$$|\varphi(y_1) - \varphi(y_2)| \leqslant L\|y_1 - y_2\|, y_1, y_2 \in D, 0 < L < \infty,$$

is accepted as such a limitation.

There are various methods for solving the considered multidimensional problem directly [11, 17] as well as efficient methods for solving the univariate problems [12, 20]. This paper considers a one-dimensional method, which is applied jointly with the dimensionality reduction scheme. The use of space filling curve (or *evolvent*) $y(x)$, where

$$\{y \in \mathbf{R}^N : -2^{-1} \leqslant y_i \leqslant 2^{-1}, 1 \leqslant i \leqslant N\} = \{y(x) : 0 \leqslant x \leqslant 1\}, \qquad (2)$$

is a well-known scheme of dimensionality reduction of the initial problem for the global optimization algorithms [18]. A mapping of the type (2) allows reducing a multidimensional problem to a univariate one at the expense of worsening its properties. In particular, the one-dimensional function $\varphi(y(x))$ is not a Lipschitz function but a Holder one:

$$|\varphi(y(x_1)) - \varphi(y(x_2))| \leqslant H|x_1 - x_2|^{\frac{1}{N}}, \; x_1, x_2 \in [0, 1],$$

where the Hölder constant H is related to the Lipschitz constant L as

$$H = 2L\sqrt{N + 3}.$$

The feasible domain can also be defined by the constraints that essentially complicate the problem. The problem statement in this case will take the following form:

$$\varphi(y^*) = \min\{\varphi(y) : y \in G\}, \; G = D \cap \{y : g_j(y) \leqslant 0, 1 \leqslant j \leqslant m\}. \qquad (3)$$

Let us set $g_{m+1}(y) = \varphi(y)$. Hereafter, we shall assume all functions $g_k(y), 1 \leqslant k \leqslant m + 1$, to satisfy the Lipschitz condition on the hyperinterval D.

Further, let us consider solving a series of q problems of the kind (3):

$$\min\{\varphi_1(y), y \in G_1\}, \min\{\varphi_2(y), y \in G_2\}, ..., \min\{\varphi_q(y), y \in G_q\}. \qquad (4)$$

Similar to [3], this work aims to formulate a method that will ensure a uniform convergence of the solutions of all problems in the series. Here and below, a uniform convergence implies a proportional decrease of the distance between the best iteration point and the global optimizer in all the problems in the series:

$$\exists \varepsilon > 0 : \forall s > 1, \forall i, j \in \{1, \ldots, q\} \; \frac{\|\tilde{y}_i(s)^* - y_i *\|_\infty}{\|\tilde{y}_j(s)^* - y_j^*\|_\infty} \leqslant \varepsilon, \qquad (5)$$

where s is the number of steps of the optimization method, $\tilde{y}_i(s)^*$ is the best iteration point in the problem i from the set (4) at the step s.

3 Description of the Global Optimization Method

Taking into account the dimensionality reduction scheme (2), let us assume that the method requires finding the global minimum of the function $\varphi(x), x \in [0, 1]$, which satisfies the Hölder condition with the constraints $g_j(x)$, which in turn satisfy this condition in the interval $[0, 1]$.

The index algorithm of global search (IAGS) for solving one-dimensional problems considered here implies construction of a sequence of points x_k, at which the values of index function z_k are calculated. The index scheme [20] was used to account for the latter. Let us assume that each function $g_i(x), 1 \leqslant i \leqslant m + 1$, is defined and computable only in the corresponding subrange $Q_i \in [a, b]$, where $\varphi(x)$ is denoted as $g_{m+1}(x)$ and

$$Q_1 = [0, 1], \; Q_{i+1} = \{x \in Q_i : g_i(x) \leqslant 0\}, \; 1 \leqslant i \leqslant m, \; Q_{m+2} = \emptyset \qquad (6)$$

Considering the definition (6) the initial problem can be rewritten as

$$\varphi(x^*) = \min\{g_{m+1}(x) : x \in Q_{m+1}\}.$$

The index scheme assigns a number called index $\nu(x)$ to each point of the search sequence x_k:

$$\nu(x) = i : x \in Q_i, x \notin Q_{i+1}, \; 1 \leqslant i \leqslant m + 1. \qquad (7)$$

In order to obtain the index of the point x we have to perform a trial defined via the following steps:

1. Sequentially compute functions $g_i(x), 1 \leqslant i \leqslant m$ until for some i^* $g_{i^*}(x) > 0$ or until $i = m + 1$. Set $\nu(x) = i^*$ or $\nu(x) = m + 1$ if $g_i(x) \leqslant 0, 1 \leqslant i \leqslant m$. Therefore, $\nu(x)$ is the number of the first violated constraint at the point x or $m + 1$ if all the constraints are satisfied at x.
2. Return pair $\nu(x), z = g_{\nu(x)}(x)$ as a result of the trial.

This approach to trials allows reducing the initial problem with functional constraints to an unconstrained problem of minimization of a discontinuous function:

$$\psi(x^*) = \min_{x \in [0,1]} \psi(x),$$

$$\psi(x) = \begin{cases} g_\nu(x)/H_\nu, & \nu < M, \\ (g_M(x) - g_M^*)/H_M, & \nu = M. \end{cases}$$

Here $M = \max\{\nu(x) : x \in [0, 1]\}$ and $g_M^* = \min\{g_M(x) : x \in Q_M\}$. Because of the definition of the number M, the problem of finding g_M^* always has a solution. If $M = m + 1$, then $g_M^* = \varphi(x^*)$. The function $\psi(x)$ satisfies the Hölder condition on the set Q_1 with the constant 1, and $\psi(x)$ can have jump discontinuities at the boundaries of the sets Q_i, $1 \leqslant i \leqslant m + 1$. Despite the values of the Hölder constants H_k and the value g_M^* are not known in advance, they can be estimated in the course of solving the problem. A set of triples $\{(x_k, \nu_k, z_k)\}, 1 \leqslant k \leqslant n$, constitutes the search information accumulated by the method upon execution of n steps.

At the first iteration of the method, the trial is performed in the arbitrary internal point x_1 of the interval $[0,1]$. The indices of the points 0 and 1 are considered to be zero indices, the values z at these points are undefined. Let us assume that $k \geqslant 1$ iterations of the method have been performed. In the course of this performance the trials were conducted at k points $x_i, 1 \leqslant i \leqslant k$. Then, the points x^{k+1} of the search trials of the next $(k+1)^{\text{th}}$ iteration are defined according to the rules:

Step 1. Reassign lower indices to the points of the set $X_k = \{x^1, \ldots, x^k\} \cup \{0\} \cup \{1\}$, which includes the boundary points of the interval $[0,1]$ as well as the points of preceding trials in the order of increasing coordinate values, i.e.

$$0 = x_0 < x_1 < \ldots < x_{k+1} = 1, \tag{8}$$

and compare them with the values $z_i = g_\nu(x_i), \nu = \nu(x_i), i = \overline{1,k}$ computed at these points.

Step 2. For each integer number $\nu, 1 \leqslant \nu \leqslant m+1$, determine the corresponding set I_ν of the lower indices of the points, at which the values of the functions $g_\nu(x)$ are computed:

$$I_\nu = \{i : \nu(x_i) = \nu, 1 \leqslant i \leqslant k\}, 1 \leq \nu \leq m+1,$$

and determine the maximum value of the index $M = \max\{\nu(x_i), 1 \leq i \leq k\}$.

Step 3. Compute current estimate for the unknown Hölder constant:

$$\mu_\nu = \max \left\{ \frac{|g_\nu(x_i) - g_\nu(x_j)|}{(x_i - x_j)^{\frac{1}{N}}} : i, j \in I_\nu, i > j \right\}. \tag{9}$$

If the set I_ν contains less than two elements or if the value μ_ν is equal to zero, then assume $\mu_\nu = 1$.

Step 4. For all the nonempty sets $I_\nu, \nu = \overline{1,M}$ compute the estimates

$$z_\nu^* = \begin{cases} \min\{g_\nu(x_i) : x_i \in I_\nu\}, & \nu = M, \\ 0, & \nu < M. \end{cases} \tag{10}$$

Step 5. For each interval $(x_{i-1}, x_i), 1 \leqslant i \leqslant k$ compute the characteristic

$$R(i) = \begin{cases} \Delta_i + \frac{(z_i - z_{i-1})^2}{(r_\nu \mu_\nu)^2 \Delta_i} - 2\frac{z_i + z_{i-1} - 2z_\nu^*}{r_\nu \mu_\nu}, & \nu = \nu(x_i) = \nu(x_{i-1}), \\ 2\Delta_i - 4\frac{z_{i-1} - z_\nu^*}{r_\nu \mu_\nu}, & \nu = \nu(x_{i-1}) > \nu(x_i), \\ 2\Delta_i - 4\frac{z_i - z_\nu^*}{r_\nu \mu_\nu}, & \nu = \nu(x_i) > \nu(x_{i-1}), \end{cases} \tag{11}$$

where $\Delta_i = (x_i - x_{i-1})^{\frac{1}{N}}$. The values $r_\nu > 1, \nu = \overline{1,m}$ are the parameters of the algorithm. They define the products $r_\nu \mu_\nu$ used in computing the characteristics as the estimates of the unknown Hölder constants.

Step 6. Select the maximum characteristic:

$$t = \arg \max_{1 \leqslant i \leqslant k+1} R(i). \tag{12}$$

Step 7. Perform the next trial in the middle of the interval (x_{t-1}, x_t) if the indices of its end points are not the same: $x^{k+1} = \frac{1}{2}(x_t + x_{t-1})$. Otherwise, perform the trial at the point

$$x^{k+1} = \frac{1}{2}(x_t + x_{t-1}) - \operatorname{sgn}(z_t - z_{t-1}) \frac{|z_t - z_{t-1}|^N}{2 r_\nu \mu_\nu^N}, \nu = \nu(x_t) = \nu(x_{t-1}),$$

and then increment k by 1.

The algorithm stops if the condition $\Delta_t \leqslant \varepsilon$ is satisfied. Here $\varepsilon > 0$ is a predefined precision. The values

$$\varphi_k^* = \min_{1 \leqslant i \leqslant k} \varphi(x_i), \quad x_k^* = \arg \min_{1 \leqslant i \leqslant k} \varphi(x_i) \tag{13}$$

are assumed as the estimates of the global solution.

Next, following the approach described in [3], we shall use q copies of IAGS working synchronously to solve the problem series (4). The only difference is that when selecting the interval with the maximum characteristic at Step 6, the choice will be made from all intervals, which were generated by q copies of IAGS. If the maximum characteristic corresponds to the ith problem, then Step 7 is performed in the ith copy of the method while the other copies stay idle. This way, at every iteration the trial is performed for the most promising problem in terms of the characteristics (11). This allows distributing the resources of the method among the problems dynamically. Let's denote this method as MIAGS.

The parallel modification of the method does not differ from the one considered in [3] and consists in the selection of p intervals at Step 6 and performing p trials in parallel at the following step. All resources of the method within the framework of the iteration may be focused on a single problem as well as on $l \leqslant p$ problems simultaneously (depending on the problem, to which the intervals selected by the method belong).

3.1 Convergence Conditions

The conditions of convergence of the method described in Section 3 in case of $q = 1$ are given in [20].

Theorem 1. *(Sufficient convergence conditions) Let us assume that the following conditions are true:*

1. *$D \neq \emptyset$, the problem (3) has a solution.*
2. *Functions $g_j(y) \leqslant 0, 1 \leqslant j \leqslant m + 1$, are Lipschitz functions with respective constants L_i over the domain D (here $g_{m+1}(y) = \varphi(y)$).*
3. *If k from (8) is sufficiently large, the values μ_ν from (9) satisfy the inequalities*

$$r_\nu \mu_\nu > 2^{3-1/N} L_\nu \sqrt{N + 3}, \ 1 \leqslant \nu \leqslant m + 1. \tag{14}$$

Then any limit point $\overline{y}$ of the sequence $\{y_k\} = \{y(x_k)\}$ generated by the index method, the problem (3) is feasible and satisfies the conditions

$$\varphi(\overline{y}) = \inf\{\varphi(y^k) : g_i(y^k) \leqslant 0, 1 \leqslant i \leqslant m, k = 1, 2, \ldots\} = \varphi(y^*). \qquad (15)$$

Remark 1. From the relationship between the Hölder and Lipschitz constants and from the condition (15) it follows that the parameters r_ν from (11) should satisfy the condition

$$r_\nu > 2^{2-1/N}. \qquad (16)$$

Theorem 2. *(On the convergence conditions of MIAGS) Let the conditions 1-3 of the Theorem 1 for each problem i, $1 \leqslant i \leqslant q$ from (4), be true i.e. each problem can be solved by IAGS. Then solving all of the q problems by MIAGS will generate q infinite sequences $\{y_i^k\}$, $1 \leqslant i \leqslant q$, such that*

$$\varphi_i(\overline{y_i}) = \inf\{\varphi(y_i^k) : g_j^i(y_i^k) \leqslant 0, 1 \leqslant j \leqslant m_i, k = 1, 2, \ldots\} = \varphi_i(y_i^*).$$

Proof. Let us consider two random problems from the set (4)

$$\begin{aligned}
\min\{\varphi(y) : y \in D_1, g_j^\varphi(y) \leqslant 0, 1 \leqslant j \leqslant m_1\}, \\
\min\{\psi(y) : y \in D_2, g_j^\psi(y) \leqslant 0, 1 \leqslant j \leqslant m_2\}.
\end{aligned} \qquad (17)$$

Let us denote the characteristics (11) for the first problem as $R_\varphi(i)$ and for the second problem as $R_\psi(j)$. Considering that, we have:

$$\begin{aligned}
R_\varphi(t_\varphi) = \max_{1 \leqslant i \leqslant k} R_\varphi(i), \\
R_\psi(t_\psi) = \max_{1 \leqslant j \leqslant s} R_\psi(j),
\end{aligned} \qquad (18)$$

where k corresponds to the number of trials in the first problem and s corresponds to the number of trials in the second problem. The sequence of trials $\{v^k\}$ corresponds to the first problem and the sequence of trials $\{u^s\}$ corresponds to the second problem. The values $z^k = g_\nu^\varphi(v_k), \nu = \nu(v_k)$ correspond to the trial points $\{v^k\}$, and the values $w^s = g_\nu^\psi(u_s), \nu = \nu(u_s)$ correspond to the trial points $\{u^s\}$.

When the two problems are solved simultaneously, the algorithm selects an interval for the next trial according to the condition:

$$R(t) = \max\{R_\varphi(i), R_\psi(j)\}. \qquad (19)$$

Let the algorithm solve two problems and the iterations counter be $l = k + s, l = 0, 1, 2, \ldots$. Then, since Theorem 1 is true for all the problems, at least one of the sequences $\{v^k\}$ and $\{u^s\}$ will be infinite (let it be $\{v^k\}$). If we can prove that both sequences are infinite, this will demonstrate the convergence in both considered problems.

Let us take a limit point $\overline{v} \in [v_{i-1}, v_i]$, where $i = i(k)$. The indices v_{i-1}, v_i can be equal or different, but, because of convergence, if k is large enough they will be stable. In the first case the algorithm will use the first branch of the rule (11), otherwise it will use one of the two other branches.

Let us consider the first case: from (9)

$$\frac{|z_i - z_{i-1}|}{\Delta_i} \leqslant \mu_{\varphi,\nu}.$$

Having considered that we can establish an upper bound:

$$\frac{(z_i - z_{i-1})^2}{r_\nu^2 \mu_{\varphi,\nu}^2 \Delta_i} = \frac{(z_i - z_{i-1})^2 \Delta_i}{(r_\nu \mu_{\varphi,\nu} \Delta_i)^2} \leqslant \frac{\Delta_i}{r_\nu^2}.$$

Therefore using the first branch of form rule (11) we get an inequality

$$R_\varphi(i) \leqslant \Delta_i (1 + \frac{1}{r_\nu^2}) - \frac{2(z_i + z_{i-1} - 2z_\nu^*)}{r_\nu \mu_{\varphi,\nu}}. \tag{20}$$

Because $\bar{v}$ is the limit point of the sequence $\{v^k\}$ and $\varphi(y(\bar{v})) \leqslant z_\nu^*, \nu = m+1$ or $z_\nu^* = 0, \nu < m+1$ and $z_{i-1}, z_i \to 0$ at $k \to \infty$:

$$\Delta_i \to 0, z_i + z_{i-1} - 2z_\nu^* \to 0. \tag{21}$$

In the second case (when one of the other two branches of the rule (11) is applied) we have

$$R_\varphi(i) = 2\Delta_i - 4\frac{z_i - z_\nu^*}{r_\nu \mu_{\varphi,\nu}}.$$

If $z_\nu^* \neq 0$ then $z_i - z_\nu^* \geqslant 0$ and

$$R_\varphi(i) = 2\Delta_i - 4\frac{z_i - z_\nu^*}{r_\nu \mu_{\varphi,\nu}} \leqslant 2\Delta_i. \tag{22}$$

Otherwise because $\bar{v}$ is a feasible point $z_i \to 0$ at $k \to \infty$.

From (20) (21) (22) for any small $\delta > 0$ there exists a large value of k such that

$$R_\varphi(i) \leqslant \delta. \tag{23}$$

Let $\alpha = \max\{\nu(u) : u \in \{u^s\}\}$. Because α is currently the highest index in the search sequence $\{u^s\}$ and according to the rule (10), $\exists j : w_\alpha^* = w_j$.

If $\nu(w_{j-1}) = \nu(w_j) = \alpha$ then the first branch of the rule (11) is applied and

$$R_\psi(j) = \Delta_j + \frac{(w_j - w_{j-1})^2}{r_\alpha^2 \mu_{\psi,\alpha}^2 \Delta_j} - \frac{2(w_j + w_{j-1} - 2w_\alpha^*)}{r_\alpha \mu_{\psi,\alpha}} \geqslant$$
$$\geqslant \Delta_j - \frac{2(w_j + w_{j-1} - 2w_\alpha^*)}{r_\alpha \mu_{\psi,\alpha}} = \Delta_j - \frac{2\Delta_j(w_{j-1} - w_j)}{r_\alpha \mu_{\psi,\alpha} \Delta_j} \geqslant \tag{24}$$
$$\geqslant \Delta_j - \frac{2\Delta_j}{r_\alpha} = \Delta_j \left(1 - \frac{2}{r_\alpha}\right).$$

If $\nu(w_{j-1}) \neq \nu(w_j) = \alpha$ then $\nu(w_{j-1}) < \alpha$ and the third branch of the rule (11) is applied:

$$R_\psi(j) = 2\Delta_j - 4\frac{w_j - w_\alpha^*}{r_\alpha \mu_{\psi,\alpha}} = 2\Delta_j > 0. \tag{25}$$

Taking into account the Remark 1, (24), (25) we can conclude that such an interval exists that $R_\psi(j) > 0$. At the same time, (23) is true and the inequality

$$R_\psi(j) > R_\varphi(i)$$

will be true when the value k is large enough. Thus the next scheduled iteration is performed for the second problem with objective $\psi(y)$, i.e. sequence $\{v^s\}$ will be infinite as well.

Since we considered two arbitrary problems from the given set of q problems, the theorem is true for any pair of problems from the set. By induction, the theorem is also true for the whole set.   □

4   Results of Numerical Experiments

The use of set of test problems with known solutions generated by a random mechanisms is one of the commonly accepted approaches to the comparison of optimization algorithms [4]. Experiments presented in this paper were based on two generators of problems of different nature [8,10] were used. These generators produce problems without nonlinear constraints. Therefore, the GCGen[1] [9] system was to be used to supplement these generators. This system allows generating problems with constraints based on arbitrary nonlinear functions.

The GCGen system comes with the examples of its application and construction of sets of problems each consisting of an objective function and two constraints generated by F_{GR} [10] or GKLS [8] generator. GKLS [8] allows obtaining the functions of predefined dimensionality and with predefined number of global optimums. In combination with GCGen, the following sets of problems were generated:

- 100 2-dimensional problems with two constraints;
- 100 3-dimensional problems with two constraints
- 20 4-dimensional problems with two constraints;
- a mixed class consisting of 50 problems with two-dimensional GKLS functions and 50 problems with F_{GR}. This set is generated in order to demonstrate that the efficiency of the method is preserved at essentially varying properties of the problems.

Examples of contour plots of the considered problems are presented in Fig. 1. The feasible domain is highlighted.

A test problem was considered to be solved if the optimization method executes the next trial y^k in the δ-vicinity of the global minimizer y^*, i.e. $\left\|y^k - y^*\right\| \leqslant \delta = 0.01 \left\|b - a\right\|$, where a and b are the left and right boundaries of the hypercube from (1). If this relation is not fulfilled after the limit of number of iterations had been reached, the problem was considered to be not solved.

---

[1] The original code of the system can be accessed at https://github.com/UNN-ITMM-Software/GCGen

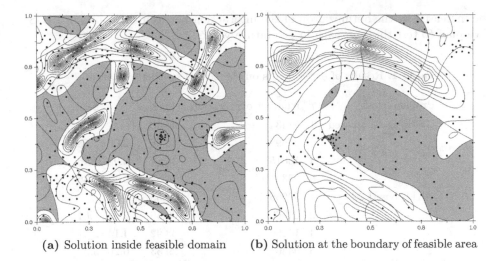

(a) Solution inside feasible domain   (b) Solution at the boundary of feasible area

Fig. 1. Contour plots and trial points of IAGS in two synthetic problems

When evaluating the quality of the method and its implementation, besides increased computation speed due to parallelization, we shall also account for the mean maximum distance (in terms of l_∞-norm) from the current estimate of the optimum to its actual position computed on the set of problems (4): D_{avg} and D_{max}. The dynamics of these magnitudes in the course of the optimization shows how uniformly the method distributes the resources among the problems.

The parallel method was implemented in C++ with the use OpenMP technology for parallelization of the trial execution process in the shared memory. All numerical experiments were carried out using a computer with the following configuration: Intel Core i7-7800X, 64 GB RAM, Ubuntu 16.04 OS, GCC 5.5 compiler.

4.1   Solution Results of Generated Problems

The solution results of test problems by the sequential and parallel versions of the MIAGS are presented in Table 1. For all two-dimensional classes of problems, the parameter $r = 4.7$. In the case of the three and four-dimensional problems, $r = 4.7$. The convergence speed was improved by applying ε-reservation technique from [20] Chapter 8.3 with $\varepsilon = 0.1$. In all the experiments, an additional computational load was introduced into the objective function and constraints to get the duration of a single call of a problem function equal to approximately 1 ms.

Table 1 shows that the speedup in the iterations $S_i = \frac{iters(p=1)}{iters(p=i)}$ increased linearly with increasing number of threads p, whereas the speed of calculations $S_t = \frac{time(p=1)}{time(p=i)}$ increased at a lower rate, which points to a non-ideal implementation of the algorithm. The actual speedup, the upper limit for which is S_i, can be increased by the optimization of the interaction between the copies of IAGS. We plan to test this approach in our future works.

Table 1. Results of experiments on the sets of synthetic problems

Problem class	p	Number of iterations	Time, s	S_i	S_t
GKLS & F_{GR} based	1	51434	90.20	-	-
	2	25698	56.96	2.00	1.58
	4	13015	36.67	3.95	2.46
	6	8332	26.85	6.17	3.36
GKLS based 2d	1	59066	97.53	-	-
	2	29060	60.56	2.04	1.61
	4	14266	38.92	4.14	2.51
	6	9436	29.53	6.26	3.30
GKLS based 3d	1	782544	1117.55	-	-
	2	397565	752.92	1.97	1.48
	4	208073	526.67	3.76	2.12
	6	142089	445.45	5.50	2.51
GKLS based 4d	1	14021720	15806.6	-	-
	2	6313070	7254.85	2.22	2.18
	4	3479344	4932.55	4.03	3.20
	6	2783339	3955.38	5.04	3.99

In order to demonstrate the uniform convergence of MIAGS, all test problems have been solved by IAGS as well. IAGS is comparable with other stochastic and deterministic derivative-free global optimization algorithms [19]. Thus, it provides a strong baseline in solving a single global optimization problem. Figure 2 presents the plots of mean and maximum distances from the actual optima to the current estimates of the optima when solving a series of problems generated by two different generators separately (solid line) and jontly (dashed line). Despite the essential differences in the structure of problems, the MIAGS decreased the maximum deviations of the estimates from the optima, as well as the mean optima much faster. It evidences that the uniform convergence over the whole set of problems has been achieved. In the case of the sequential solving of the problems, the magnitude D_{max} takes its maximum value until the last problem is solved.

4.2   Example of Solving a Multicriteria Problem

In order to demonstrate the efficiency of the approach in the balancing of the load, let us consider an example, in which a set of problems of the kind (4)

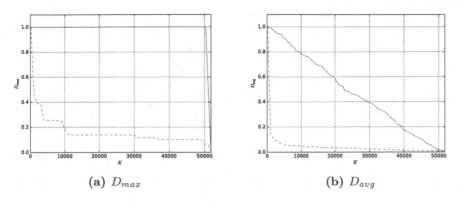

(a) D_{max} (b) D_{avg}

Fig. 2. Dynamics of D_{avg} and D_{max} in the course of solving a set of the two- dimensional problems generated by two different generators GKLS and F_{GR}

is generated as a result of scalarization of a multicriteria optimization problem with constrains.

Let us consider a test problem from [21]:

$$Minimize \begin{cases} f_1(y) = 4y_1^2 + 4y_2^2 \\ f_2(y) = (y_1 - 5)^2 + (y_2 - 5)^2 \end{cases}, y_1 \in [-1, 2], y_2 \in [-2, 1],$$

s.t.   (26)

$$\begin{cases} g_1(y) = (y_1 - 5)^2 + y_2^2 - 25 \leqslant 0, \\ g_2(y) = -(y_1 - 8)^2 - (y_2 + 3)^2 + 7.7 \leqslant 0. \end{cases}$$

Let us use the Germeyer convolution for the scalarization of the problem (26). After the convolution, the scalar objective function takes the form:

$$\varphi(y, \lambda_1, \lambda_2) = \max\{\lambda_1 f_1(y), \lambda_2 f_2(y)\},$$ (27)

where $\lambda_1, \lambda_2 \in [0, 1]$, $\lambda_1 + \lambda_2 = 1$. Testing all possible convolution coefficients allows finding the whole set of Pareto-optimal solutions of the problem (26). For the numerical construction of the Pareto set, let us select 100 sets of coefficients (λ_1, λ_2) so that $\lambda_1^i = ih$, $\lambda_2^i = 1 - \lambda_1^i$, $h = 10^{-2}$, $i = \overline{1,100}$.

Computational resources were limited to 2500 trials. The set of auxiliary scalar problems was solved by two methods:

– each problem was solved separately using IAGS with a preset limit of 25 trials. This way, the computational resources were distributed among the problems uniformly;
– all problems were solved simultaneously using MIAGS with a preset limit of 2500 trials.

In both cases, the parameter $r = 4$.

The plots of solutions obtained by each method are presented in Fig. 3a and Fig. 3b. All plots agree with the ones presented in [21] qualitatively (the authors did not provide any other information to compare). The Pareto curve in Fig. 3a

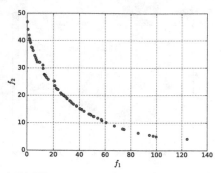
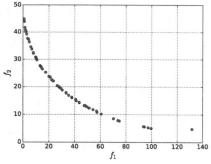

(a) IAGS, separate solving of the problems   (b) MIAGS for the set of problems

Fig. 3. Numerical estimates of Pareto set in the problem (26), obtained after 2500 trials

has concavities that do not match the solution presented in [21], which means there are not enough resources for solving some auxiliary problems. To evaluate the quality of solution, the index $Spacing(SP)$ [15] featuring the density of the points approximating the Pareto set was computed.

$$SP(S) = \sqrt{\frac{1}{|S|-1}\sum_{i=1}^{|S|}(\overline{d} - d_i)^2}, \; \overline{d} = mean\{d_i\},$$
$$d_i = \min_{s_i, s_j \in S: s_i \neq s_j} ||F(s_i) - F(s_j)||_1, \; F = (f_1, f_2).$$

When problems were solved separately $SP_{single} = 0.984$, when the load balancing method was applied $SP_{multi} = 0.749$, which evidences a better quality of the approximation of the solution.

5   Conclusion

This paper demonstrates how the support of the non-convex constraints can be implemented in the algorithm to solve a set of the global optimization problems. This study allowed to find the sufficient conditions of convergence for the developed method. The numerical experiments conducted within this research demonstrate the advantages of the considered approach over separate solution of the problems. The efficiency of joint solution of a set of problems was demonstrated on the example a multicriteria problem with nonlinear constraints. Further research in this direction should improve current implementation of the algorithm by reducing the support costs of the search information for the set of problems. This, in turn, should improve the calculation speed due to parallelization. There are also plans to implement a version of considered algorithm in the distributed memory according to the scheme described in [2].

References

1. Barkalov, K., Lebedev, I.: Comparing two approaches for solving constrained global optimization problems. In: Battiti, R., Kvasov, D.E., Sergeyev, Y.D. (eds.) LION 2017. LNCS, vol. 10556, pp. 301–306. Springer, Cham (2017). https://doi.org/10. 1007/978-3-319-69404-7_22

2. Barkalov, K., Lebedev, I.: Parallel algorithm for solving constrained global optimization problems. In: Malyshkin, V. (ed.) PaCT 2017. LNCS, vol. 10421, pp. 396–404. Springer, Cham (2017). https://doi.org/10.1007/978-3-319-62932-2_38

3. Barkalov, K., Strongin, R.: Solving a set of global optimization problems by the parallel technique with uniform convergence. J. Global Optim. **71**(1), 21–36 (2017). https://doi.org/10.1007/s10898-017-0555-4

4. Beiranvand, V., Hare, W., Lucet, Y.: Best practices for comparing optimization algorithms. Optim. Eng. **18**(4), 815–848 (2017). https://doi.org/10.1007/s11081-017-9366-1

5. Dostl, Z.: Optimal Quadratic Programming Algorithms: With Applications to Variational Inequalities, 1st edn. Springer Publishing Company Incorporated (2009)

6. Ehrgott, M.: Multicriteria Optimization. Springer, Heidelberg (2005). https://doi. org/10.1007/3-540-27659-9

7. Evtushenko, Y., Posypkin, M.: A deterministic approach to global box-constrained optimization. Optim. Lett. **7**, 819–829 (2013). https://doi.org/10.1007/s11590-012-0452-1

8. Gaviano, M., Kvasov, D.E., Lera, D., Sergeev, Y.D.: Software for generation of classes of test functions with known local and global minima for global optimization. ACM Trans. Math. Softw. **29**(4), 469–480 (2003). https://doi.org/10.1145/962437.962444

9. Gergel, V., Barkalov, K., Lebedev, I., Rachinskaya, M., Sysoyev, A.: A flexible generator of constrained global optimization test problems, vol. 2070, p. 020009, February 2019. https://doi.org/10.1063/1.5089976

10. Grishagin, V.A.: Operating characteristics of some global search algorithms (in Russian). Prob. Stochast. Search **7**, 198–206 (1978)

11. Jones, D.R.: The direct global optimization algorithm. In: The Encyclopedia of Optimization, pp. 725–735. Springer, Heidelberg (2009). https://doi.org/10.1007/978-0-387-74759-0_128

12. Norkin, V.I.: Towards pijavskyj's method for solving common global optimization problem. Comp. Math. Math. Phys. **32**, 992–1006 (1992)

13. Paulavivcius, R., Zilinskas, J., Grothey, A.: Parallel branch and bound for global optimization with combination of lipschitz bounds. Optim. Method. Softw. **26**(3), 487–498 (1997). https://doi.org/10.1080/10556788.2010.551537

14. Pham Dinh, T., Le Thi, H.A.: Recent Advances in DC Programming and DCA, pp. 1–37. Springer, Heidelberg (2014). https://doi.org/10.1007/978-3-642-54455-2_1

15. Riquelme, N., Von Lucken, C., Baran, B.: Performance metrics in multi-objective optimization. In: 2015 Latin American Computing Conference (CLEI), pp. 1–11, October 2015. https://doi.org/10.1109/CLEI.2015.7360024

16. Sergeyev, Y.D., Famularo, D., Pugliese, P.: Index branch-and-bound algorithm for lipschitz univariate global optimization with multiextremal constraints. J. Glob. Optim. **21**(3), 317–341 (2001)

17. Deterministic Global Optimization. SO. Springer, New York (2017). https://doi. org/10.1007/978-1-4939-7199-2

18. Sergeyev, Y.D., Strongin, R.G., Lera, D.: Introduction to Global Optimization Exploiting Space-filling Curves. Springer, New York (2013). https://doi.org/10. 1007/978-1-4614-8042-6

19. Sovrasov, V.: Comparison of several stochastic and deterministic derivative-free global optimization algorithms. In: Khachay, M., Kochetov, Y., Pardalos, P. (eds.) MOTOR 2019. LNCS, vol. 11548, pp. 70–81. Springer, Cham (2019). https://doi. org/10.1007/978-3-030-22629-9_6

20. Strongin R.G., Sergeyev, Y.D.: Global optimization with non-convex constraints. Sequential and parallel algorithms. Kluwer Academic Publishers, Dordrecht (2000). https://doi.org/10.1007/978-1-4615-4677-1

21. To, T.B., Korn, B.: MOBES: a multiobjective evolution strategy for constrained optimization problems, March 1999

Combinatorial and Discrete Optimization

Multi-core Processor Scheduling with Respect to Data Bus Bandwidth

Anton V. Eremeev[1,3]([✉]) [ID], Anton A. Malakhov[2], Maxim A. Sakhno[1,3],
and Maria Y. Sosnovskaya[1,3]

[1] Sobolev Institute of Mathematics, Novosibirsk, Russia
eremeev@ofim.oscsbras.ru
[2] Intel Corporation, Nizhny Novgorod, Russia
[3] Dostoevsky Omsk State University, Omsk, Russia

Abstract. The paper considers the problem of scheduling software modules on a multi-core processor, taking into account the limited bandwidth of the data bus and the precedence constraints. Two problem formulations with different levels of problem-specific detail are suggested and both shown to be NP-hard. A mixed integer linear programming (MILP) model is proposed for the first problem formulation, and a greedy algorithm is developed for the second one. An experimental comparison of the results of the greedy algorithm and the MILP solutions found by CPLEX solver is carried out.

Keywords: Multi-core processor · Data bus · Scheduling · Greedy algorithm · Mixed integer linear programming

1 Introduction

The goal of the paper is to investigate resource constraint scheduling problems that arise when developing a program for a multi-core processor. In this case, it is necessary to schedule the execution of software modules on the processor cores, taking into account the restrictions on the data bus bandwidth. The data bus is a part of the system bus that is used to transfer data between computer components, in this particular case, between the CPU and the random access memory (RAM). Different software modules need different amount of data flow via data bus, therefore in the case of simultaneous execution of several modules, each one of them can take longer time than in the case of single-thread execution. The problem of scheduling software modules on a multi-core processor w.r.t the limited bandwidth of the data bus is important for processor manufacturers and parallel software developing companies, since the more efficiently the data bus is used, the higher the software performance.

From the point of view of scheduling theory, the problem of allocating the software modules to processor cores with respect to the limited data bus bandwidth is similar to the scheduling problems with renewable resources (see e.g. [12]), but unlike those problems, in our case the resource constraint (data

© Springer Nature Switzerland AG 2020
N. Olenev et al. (Eds.): OPTIMA 2020, CCIS 1340, pp. 55–69, 2020.
https://doi.org/10.1007/978-3-030-65739-0_5

bus bandwidth) does not exclude some infeasible combinations of jobs (software modules) but rather increases their execution times. A distinctive feature of our problem is that each job would be processed at different speeds depending on its requirement of the data bus bandwidth and the loading of the data bus by the simultaneous jobs on other cores.

There are a number of approaches to task scheduling with variable processing times in the literature. First of all, in the area of parallel software development for multi-core processors, such problems are usually solved using fast heuristics, which work in the online mode, i.e. the jobs arrive sequentially and only a limited number of jobs is considered in each moment. The task scheduling heuristics proposed in [10,18] and some other works are based on the principle that tasks should be allocated on the CPU cores in a complementary fashion, so that the tasks with most different resource consumption requirements are co-scheduled for simultaneous execution (in [10,18] such resources imply the usage of data bus bandwidth and the cache utilization at different levels).

The tasks scheduling method proposed in [7] is based on the co-run *degradation coefficients*, equal to an increase in the execution time of an application when it shares a cache with a co-runner, relative to running solo. In the case of dual-core CPUs, the threads may be represented as nodes connected by edges, and the weights of the edges are given by the sum of the mutual co-run degradations between the two threads. Then, under some simplifying assumptions, an optimal schedule may be found by solving a min-weight perfect matching problem. In the case of greater number of cores per CPU, the problem is shown to be NP-hard and several heuristic approximation algorithms are suggested. Although the methodology from [7] and the corresponding algorithms would be too expensive to use online, they are acceptable for offline evaluation of the quality of other approaches.

Authors of [17] propose a novel fairness-aware thread co-scheduling algorithm based on non-cooperative game to reduce L2 cache misses. The execution time of a thread varies depending on which threads are running on other cores of the same chip, because different thread combinations result in different levels of cache contention. In [16], the cache on each of m chips is shared by u cores on the each chip. The execution speed of a job running on a chip depends on what jobs are placed on the same chip. The number of jobs is equal to the total number of cores, all the jobs start at the same time. It is proved that the problem is NP-hard and a series of algorithms is presented to compute or approximate the optimal schedules.

In the production scheduling applications, the problem formulations with variable processing times are also important, e.g. in [9], a coke production scheduling problem is considered, where jobs influence on the processing time of other jobs due to increased production unit temperature. The authors of [9] construct an integer programming model to minimize the makespan, propose several heuristics, including a genetic algorithm, and compare their performance.

In the scheduling theory, similar problem formulations may be found in the area of scheduling with controllable processing times. The models and methods for the case of preemptive scheduling are surveyed in [15]. In [13], the problem of job scheduling on identical parallel machines is considered, where the processing time of jobs is controlled by allocating a non-renewable shared limited resource. It is proved that if job preemptions are allowed, then the problem of minimizing the makespan time is solvable in $O(n^2)$ operations. In the present paper, however, we consider a non-preemptive problem formulation. Since the data bus bandwidth is a renewable resource, we refer to [8] and [2] for the surveys on problems with renewable resources, where the resource allocations may vary over time. The case of discrete resources is considered in [2], and the continuous resources are considered in [8]. The latter paper contains a problem formulation similar to our formulation F1 considered below, however in [8] it is supposed that the amount of resources allocated to each job is limited, but continuous and decided by the scheduler at each moment of time. In our case, the jobs execution speeds are completely defined by the set of co-scheduled jobs on the other cores (or machines in the traditional scheduling terminology). In a recent work [1], the authors focus on assignment of shared continuous resources *to the processors*, while the job assignment to processors and the ordering of the jobs is fixed. These are the main differences to the problem considered in our paper. One more difference is that unlike [1], we make a continuous time assumption. The authors of [1] show that, even for unit size jobs, finding an optimal solution is NP-hard if the number of processors is a part of the input, however a polynomial-time algorithm for any constant number of processors and unit size jobs exists.

In the present paper, two mathematical problem formulations for the problem of allocating the software modules to processor cores are proposed with different levels of problem-specific detail and both shown to be NP-hard. A mixed integer linear programming (MILP) model using the concept of event points (see e.g. [4]) is proposed for the more detailed problem formulation, and a greedy algorithm is developed for the other one. A comparison of the greedy algorithm results and the MILP solutions found by CPLEX solver is carried out.

The paper has the following structure. Two problem formulations are proposed in Sect. 2. NP-hardness of both problem formulations is shown in Sect. 3. A mixed integer linear programming model for the first problem formulation is suggested in Sect. 4. The greedy heuristic for the second problem formulation is described in Sect. 5. Methods of real-life input data generation and testing are explained in Sect. 6. The results of computational experiments are presented in Sect. 7. Concluding remarks are given in Sect. 8.

2   Problem Formulations

Informally, our problem is to schedule execution of software modules (jobs) on a number of processor cores, while there is one resource of a renewable type, the bandwidth of the data bus, and the precedence constraints for execution of these modules are given as a partial order on the set of jobs, and the objective is to minimize the makespan. Here we assume that each module creates a uniform data flow through the data bus, so that the amount of information sent by a module through the data bus in both directions (from CPU to RAM and back) is proportional to the fraction of the completed job (i.e. the ratio of the executed elementary operations of a job to the total number of elementary operations in this job). Examples of software modules with (almost) uniform data flow may be the computational routines with multiple repetitions of the same loop or copying large data arrays.

Formulation F1. There are m jobs, c processor cores. The jobs are performed with no preemption and do not migrate from one core to another during the execution. No more than one job can be performed on a single core.

For each job p, $p = 1, ..., m$, let s_p denote its processing time under ideal conditions. Here and below, by *ideal conditions* we mean job execution when no other job is performed simultaneously.

We will call *a configuration* any set of jobs which may be performed simultaneously on different cores, taking into account the partial order on the set of jobs (a configuration can not contain a pair of jobs where one job precedes another according to the partial order) and the restrictions on the number of cores. Let K denote the set of all configurations. Suppose that in zero configuration no job is performed. Clearly, the partial order on the set of jobs also induces a partial order on the set of configurations K.

Let us call *a processing speed* of job p in configuration $k \in K$ the ratio of the time of full execution of job p under ideal conditions to the time of full execution of job p, if p was executed all this time in configuration k. Throughout each configuration, the speed of all jobs is supposed to be constant, but the processing speed of a job may vary during its execution, depending on the configuration in which it is performed. The configuration can be changed in two cases: the first case is when one of the jobs in the current configuration has completely completed and the second case is when some job(s) is added to the current configuration. If the configuration is changed, the speed of those jobs that are still in progress may change.

So, for each configuration of $k \in K$ we know which jobs it consists of. For each job p in configuration k, the speed of its execution v_{pk} is known.

The problem consists in scheduling the jobs on the processor cores with the minimum makespan (i.e. the time of completion of all jobs).

Since the number of configurations can be very large (up to $\sum_{i=0}^{c} \binom{m}{i}$, depending on the partial order), the problem formulation F1 can be simplified by introducing the assumption that the job execution speeds are calculated based on their actual consumption of the data bus. In practice, the job speed depends on a large number of factors such as the number of memory access channels for the processor, the number of processor cache levels and their free volume, the processor frequency and its temperature (depending on the specific processor and related components). Explicit consideration of all these factors is beyond the scope of this paper. Based on practice, we suggest another problem formulation which is based on the jobs usage of data bus bandwidth. This problem formulation can be written as follows.

Formulation F2. There are m jobs, c processor cores and one renewable resource, the data bus. Just like in Formulation F1, the jobs are performed without preemption or migration from one core to another, and a partial order on jobs is given. It is required to schedule the jobs on the processor cores with the minimum makespan.

Now we suppose that for each job p, $p = 1, ..., m$, the percentage of data bus consumption b_p under ideal conditions is known. During the execution of job p, a smaller percentage of the data bus can be allocated than b_p if other jobs are simultaneously performed on other cores. Denote by z_{pk} the actually allocated percentage of the data bus to job p in configuration k. In practice, the distribution of values z_{pk} among the threads is very hardware-specific and depends on many factors, which we can not afford to take into account (see e.g. [11]). As a simple approximation, we assume that the data bus bandwidth allocation to jobs in any configuration k may be found by Algorithm 1, described below. The speed v_{pk} of execution of job p in the configuration k is then proportional to the ratio z_{pk}/b_p.

Algorithm 1. Calculation of data bus consumption for a given configuration

Step 0. Put the percentage of the free data bus $freePercent := 100\%$ (the entire data bus is free) and set the number of jobs for which the data bus is not allocated, $jobsCount$ to be the number of jobs in configuration k.

Step 1. While $jobsCount$ is not 0, do:

    1.1 Calculate a percentage of data bus that can be allocated to each job:
$$percent := freePercent/jobsCount.$$

    1.2 If the configuration has such a job p that $b_p < percent$, then put:

        $z_{pk} := b_p,$
        $freePercent := freePercent - b_p,$
        $jobsCount := jobsCount - 1.$

    If no such job is found, then put $z_{pk} := percent$ for each remaining job and $jobsCount := 0$.

Step 3. Output computed values z_{pk}.

This method of capacity allocation is different from the concurrent network flow allocation, well-known in multicommodity flow problems (see e.g. [14]), where the ratio of the flow of each commodity to the predefined flow demand for that commodity must be the same for all commodities. We expect that the capacity allocation represented by Algorithm 1 is more adequate to the case of data bus information flows because in this case a software module has no explicit way to communicate its flow demand to the system.

3 Problem Complexity

We will show that the decision versions in both formulations contain an NP-complete special case of MULTIPROCESSOR SCHEDULING problem [3] as a special case. Here is a formulation of this problem:

Given a set T of tasks, number $w \in \mathbb{Z}^+$ of processors, length $l(t) \in \mathbb{Z}^+$ for each $t \in T$, and a deadline $D \in \mathbb{Z}^+$, is there a w-processor schedule for T that meets the overall deadline D?

In [3] it is also proved that the MULTIPROCESSOR SCHEDULING problem remains NP-complete in the special case of $w = 2$.

Proposition 1. *The problem of multi-core processor scheduling with respect to data bus bandwidth is NP-hard for both formulations F1 and F2.*

Proof. The MULTIPROCESSOR SCHEDULING problem is a special case of the decision version of the problem of multi-core processor scheduling with respect to data bus bandwidth in Formulation F2 in the special case where: (i) jobs do not slow each other, (ii) there is no partial order constraint, and (iii) the set of tasks T is equal to the set of jobs, assuming that the number of processors is the number of cores.

To prove the NP-hardness of the problem in Formulation F1, put the number of cores equal to 2. In this case, the number of configurations is $1 + m + \frac{m(m-1)}{2}$ and, therefore, the input size of the problem in question is limited by a polynomial of the input size of the MULTIPROCESSOR SCHEDULING problem.   □

4   Mixed Integer Linear Programming Model

Consider a mixed integer linear programming (MILP) model for the first problem formulation. We define the concept of an event point similar to that introduced in [4]. In this paper, an event point characterises a time interval in which a single configuration is performed. It is defined by the number of the interval, its duration and the configuration used in it.

Let $P = \{1, ..., m\}$ denote the set of all jobs. The following set of parameters may be computed on the basis of an instance given in Formulation F1:

- $q_{pk} = 1$ if and only if job p is performed in configuration k, 0 otherwise.
- $a_{ij} = 1$ if and only if configuration i should run after configuration j, 0 otherwise.
- T_{max} is an upper bound on the duration of any configuration at any event point.

Let us denote by $N = \{0, 1, 2, ..., e\}$ the set of all event points, where e is the maximal index of event points, and introduce the problem variables:

- t_{nk} is duration of execution of configuration k at the event point n.
- $d_{nk} = 1$ if and only if configuration k is performed at the event point n, 0 otherwise. For consistency of the MILP model, we assume that the zero configuration is performed at the zero point of events. Then $d_{00} = 1$ and $d_{0k} = 0$, $k \in K$.
- $y_{pn} = 1$ if and only if job p started at the event point n, 0 otherwise.

Then the MILP model can be written as follows:

$$\min \sum_{n \in N} \sum_{k \in K} t_{nk}, \tag{1}$$

$$t_{nk} \geq 0, \ n \in N, \ k \in K, \tag{2}$$

$$t_{nk} \leq d_{nk} T_{max}, \ n \in N, k \in K, \tag{3}$$

$$\sum_{k \in K} d_{nk} = 1, \ n \in N, \tag{4}$$

$$\sum_{n \in N} \sum_{k \in K} t_{nk} v_{pk} = s_p, \ p \in P, \tag{5}$$

$$\sum_{k \in K} d_{nk} q_{pk} - \sum_{k \in K} d_{n-1,k} q_{pk} \leq y_{pn}, \ p \in P, \ n \in N, \tag{6}$$

$$\sum_{n \in N} y_{pn} = 1, \ p \in P, \tag{7}$$

$$a_{k_1,k_2} d_{n_1,k_2}(n_1 + 1) \leq a_{k_1,k_2}(d_{n_2,k_1} + (1 - d_{n_2,k_1})e)n_2, \ k_1, k_2 \in K, \ n_1, n_2 \in N, \tag{8}$$

$$d_{nk} \in \{0,1\}, \ y_{pn} \in \{0,1\}, \ p \in P, \ n \in N, \ k \in K. \tag{9}$$

The objective function (1) defines the makespan criterion. Inequality (2) guarantees that the duration of execution of configuration k at the event point n is non-negative, and inequality (3) guarantees that the duration of execution of configuration k will be zero only if this configuration is not executed at the event point n, otherwise it will be no more than T_{max}. Equality (4) means that one and only one configuration is performed at each event point, and equality (5) means that each job must be completed completely. Inequality (6) and equality (7) guarantee continuity of job. Inequality (8) sets a partial order between configurations. Expression (9) describes the range of the d_{nk} and y_{pn} variables.

Proposition 2. *There is an optimal solution to MILP model (1)–(9) using $2m + 1$ event points, which defines an optimal schedule in Formulation F1.*

Proof. Note that each event point corresponds to a change of configurations, and a change only occurs when any job (or several jobs) has begun or has ended. Suppose that at each event point only one job begins or ends, then it is easy to see that in this case $2m$ event points are needed. We also take into account that we need a zero point of events, the point at which no configuration is performed. This implies that in the case when no two jobs start and end at the same time, the number of event points is $2m + 1$. In any other case, a smaller number of event points would be required.   □

Thus, in what follows we assign $e := 2m$.

It is worth noting that the solutions to MILP model (1)–(9) prodive the information about which configurations are performed at which event point, but do not contain the distribution of jobs to the cores. For scheduling the execution of jobs on the processor cores (as Formulation F1 requires), the following Algorithm 2 is proposed which takes as input $k_1, k_2, \ldots, k_h$, configurations sorted by execution order, as well as their execution time $l_{k_1}, l_{k_2}, \ldots, l_{k_h}$ and the number of cores c. The algorithm returns the staring time u_p and the completion time f_p for each job p.

Algorithm 2. Jobs scheduling on the basis of MILP solution

Step 1. For each job p from k_1 assign a free core and set $u_p := 0$

Step 2. For each $k_i, i = 2, \ldots, h$ do:

2.1 For each job $p \in k_i \cap k_{i-1}$, keep the same core.

2.2 For each job $p \in k_{i-1} \backslash k_i$ free the core on which job p was performed, and set $f_p := \sum_{j=1}^{i-1} l_{k_i}$.

2.3 For each job $p \in k_i \backslash k_{i-1}$ assign a free core and set $u_p := \sum_{j=1}^{i-1} l_{k_i}$.

5   Greedy Algorithm

In view of the fact that the problem is NP-hard, a constructive heuristic has been proposed for formulation F2. In what follows, this heuristic is called the *greedy algorithm*, because it assigns jobs to all cores, not allowing them to stand idle, if possible. At each iteration, the algorithm selects a set of jobs (configuration) to perform, trying to select jobs so that when allocation the data bus between them, each job gets the most closest share of the data bus to the one it needs, but at the same time the maximum possible number of cores should be loaded. After selecting a configuration, the greedy algorithm determines the completion of which of the selected jobs will lead to switching the next configuration. To this end, firstly, the algorithm calculates what percentage of the data bus will be allocated to each job, and then, on the basis of these data, it determines the speed of processing the selected jobs. To give a detailed description of the greedy algorithm, let us denote by $k_1, k_2, \ldots, k_i, \ldots$ the sequence of configurations generated by the greedy algorithm, and denote by $duration_i$ the duration of the configuration k_i. Then the algorithm can be written as follows.

Algorithm 3. Greedy algorithm

Step 0. Put percentage of free data bus $freePercent := 100\%$ (the entire data bus is free); number of free cores $freeCores := c$ (all cores are free); $i := 1$; $k_i := \emptyset$; $duration_i := 0$; time remaining for job p to completion under ideal conditions $leftTime_p := s_p$; a set of all jobs that have not started yet: $jobs := \{1, \ldots, m\}$

Iteration i. Repeat Steps 1–7:

Step 1. While $freePercent > 0$ and $freeCores$ is not 0, repeat 1.1.-1.2:

    1.1. Find an admissible (not started earlier and not forbidden by the partial order) job $p \in jobs$ for which the value $|freePercent - b_p|$ is minimal. In other words, find such valid job $p \in jobs$, which has the bus requirement closest to $freePercent$. If no such job is found, then go to **Step 3**.

    1.2. Put

        $freePercent := freePercent - b_p$;
        $freeCores := freeCores - 1$;
        $jobs := jobs - \{p\}$;
        $k_i := k_i \cup \{p\}$.

Step 2. While $freeCores$ is not 0, repeat 2.1.-2.2:

    2.1. Find an admissible job $p \in jobs$ that has the lowest data bus consumption. If no valid job is found, then go to **Step 3**.

    2.2. Put

        $freeCores := freeCores - 1$;
        $jobs := jobs - \{p\}$;
        $k_i := k_i \cup \{p\}$.

Step 3. If $k_i = \emptyset$, then go to **Step 8**. Otherwise, distribute the data bus capacity between the jobs according to Algorithm 1, which gives the value z_{pk_i} – allocated percentage of the data bus to job p in configuration k_i.

Step 4. Calculate processing speed v_{pk_i} of all jobs $p \in k_i$ in configuration k_i:

    $v_{pk_i} := z_{pk_i}/b_p$.

Step 5. Determine which job will be fully completed first in the chosen configuration and set the duration of the configuration k_i equal to the duration of this job in the configuration k_i:

    $duration_i := min_{p \in k_i} \{leftTime_p/v_{pk_i}\}$.

Step 6. For all $p \in k_i$, for which $leftTime_p/v_{pk_i}$ is equal to $duration_i$, set $leftTime_p := 0$.

Step 7. Put

    $freeCores := c$;
    $freePercent := 100\%$;
    $k_{i+1} := \emptyset$.

For all $p \in k_i$, for which $leftTime_p/v_{pk_i}$ is not equal to $duration_i$, put

    $leftTime_p := leftTime_p - duration_i v_{pk_i}/s_p$;
    $k_{i+1} := k_{i+1} \cup \{p\}$;
    $i := i + 1$;
    $freePercent := freePercent - b_p$;
    $freeCores := freeCores - 1$.

Step 8. Distribute the jobs among the cores according to Algorithm 2.

It is not difficult to see that the greedy algorithm constructs a feasible schedule and may be implemented with time complexity $O(n^2)$.

6   Methods of Data Generation and Schedules Testing

All calculations described in Sects. 6 and 7 were carried out on a computer with 16 GB of RAM and Intel Core i7-8565U 1.80 GHz CPU. The operating system used was Windows 10 version 1909. The number of threads used for calculations did not exceed the number of processor cores, so the impact of other processes and the operating system itself can be considered insignificant. Turbo Boost [6] and Hyper-threading [5] options were turned off in order to be sure that the CPU temperature and other uncontrolled factors do not influence the jobs processing times. All the programs described below were implemented in C++. For the computational experiment, the following procedures taken from the Intel MKL (Math Kernel Library) are used as jobs:

- copying a vector to another vector,
- calculation of the sum of magnitudes of the vector elements,
- calculation of a vector-scalar product and adding the result to a vector,
- calculation of the QR factorization of a matrix.

The choice of such procedures is due to the fact that they consume the data bus in different ways. The input data to the procedures has different sizes (for procedures with vectors, this is the vector length, for procedures with matrices, this is the matrix size). For vectors, the dimensions from 10 to 70 million elements were used, for matrices, the dimensions varied from 1000 to 1300. Such sizes are due to the requirement that the jobs data should not be kept in the processor's cache and their durations should not be too small (otherwise large measurement errors can occur) and they should not be too large (otherwise the measurements will take too much CPU time).

Input parameters for the generator:

- The number of jobs for which a schedule needs to be made. Values used: 4, 6, 7, 8, 10 (finding optimal solutions for 11 or more jobs takes about one hour, and for 13 and more jobs the generated model size is more than 26 Gb).
- Partial order to be generated for the jobs. Values used: (i) with a trivial partial order (no dependencies between the jobs), (ii) constructed at random (With probability 0.5 we decide that a job p_1 should be performed after job p_2. To avoid cycles, only pairs of jobs (p_1, p_2) where $(p_1 > p_2)$ are considered.), (iii) a binary tree, and (iv) one-to-many-to-one.
- Number of cores. Values used: 2, 3, 4.

6.1   Calculation of the Data Bus Consumption

All data, except for the data bus consumption by each job, necessary for the greedy algorithm, was taken from the examples generated in Sect. 6: the number of time units needed for each job, the partial order, and the number of cores. To calculate the percentage of the data bus consumed by job p, we used a program that works as follows:

– Job p was started simultaneously in c copies (c is equal to the number of cores on the computer used), after that the speed s_p^* of the job was calculated as the execution time of job p under ideal conditions, divided by the execution time of job p along with $(c-1)$ copies of the same job, then $\frac{100\%}{s_p^* c_i}$ is taken as the desired data bus consumption.

– If c copies of job p running simultaneously did not slow down each other, then this job was started with $(c-1)$ copies of job g, which has the highest data bus consumption. In this case, the percentage of data bus consumption by job p can be found as follows: $100\% - (c-1)x$, where x is the percentage of the data bus required by job g, multiplied by the speed of g in this configuration. If, in this case, no job has slowed down, then job p in any configuration does not affect the speed of other jobs, therefore, the data bus consumption by job p can be set equal to 0.

6.2 Experimental Measurement of Makespan for the Constructed Schedules

Using the generator from Subsect. 6, we calculated the real speed of jobs in various configurations. However, the greedy algorithm calculates these speeds based on data on the consumption of the data bus by each job according to formulation F2. In order to understand how adequate the completion times are estimated in the MILP model using formulation F1 and in the greedy algorithm using formulation F2, and how the greedy solutions compare to the MILP solutions, a program code was written that simulated the execution of a given schedule on the processor cores.

In this testing program, threads are created in an amount equal to the number of cores of the simulated processor. Each thread is passed a job queue in the order in which they need to be executed. Before starting to perform the next job, the thread expects the completion of all previous jobs. If the core should be idle between performing two jobs, then a fictitious job is added to the queue between the corresponding jobs.

7 Computational Experiment

Schedules constructed using GAMS modeling system with MILP model (1)–(9) were tested using the program described in Subsect. 6.2. Figure 1 shows the histogram of relative deviation (in percentage) of the makespan calculated by the CPLEX package from the measured makespan. In total, 964 schedules were tested, in all of them the deviation does not exceed 11%, in 98% of them it does not exceed 10%, and in 73% it does not exceed 5%.

Schedules constructed by the greedy algorithm were also tested using the program described in Subsect. 6.2. Figure 2 shows a histogram of relative deviation (in percentage) of completion time reported by the greedy algorithm from the real completion time.

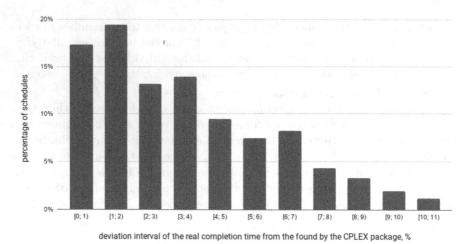

Fig. 1. Histogram of relative deviation of the minimum completion time in formulation F1 from the real completion time

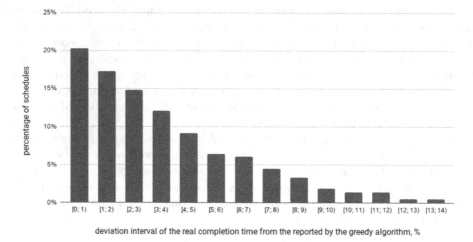

Fig. 2. Histogram of relative deviation of completion time reported by the greedy algorithm from the real completion time

In total 984 schedules with different number of jobs, different partial orders, and different number of cores were tested. In most of the cases (95%), the makespan evaluation computed in greedy algorithm differs from that obtained in the experiment by no more than 10%, and in 73% of cases by at most 5%. In 100% of cases the deviation does not exceed 14%. Such results show a fairly high accuracy of evaluation of the jobs processing time in the greedy algorithm.

Let us denote by $r := \frac{ga\_f2_{real}}{opt\_f1_{real}}$ the ratio of the real measured makespan of the greedy schedules (ga_f2_{real}) in formulation F2 to the real measured makespan of the optimal schedules (opt_f1_{real}) in formulation F1. Figure 3 shows a box-plot diagram of r ratio for different numbers of jobs.

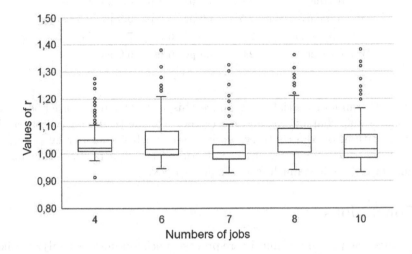

Fig. 3. Ratio of the real measured makespan for greedy schedules in formulation F2 to the measured makespan of optimal schedules in formulation F1

For each number of jobs, 192 schedules with different partial orders and number of cores were tested. It can be noted that the median ratio r for each number of jobs is close to 1.05, which allows us to conclude that the greedy algorithm is highly accurate. It is also worth noting that even in the worst cases, the makespan of greedy schedule exceeds the mistaken found by the MILP model in formulation F1 not more than by a factor of 1.4. In Fig. 3, one can also see that in some cases the solutions of the greedy algorithm in real life turn out to be faster than the optimal solutions, however, the difference does not exceed 10% and may be related to the error in calculating the input data of the problem and in testing of the obtained solutions.

Tables 2 and 1 show the CPU time of the greedy algorithm (Table 2) and the CPU time of the CPLEX package (Table 1) for different types of partial order and different numbers of jobs. The CPLEX package most quickly finds solutions

Table 1. Average CPU time of the CPLEX package

	4 jobs	6 jobs	7 jobs	8 jobs	10 jobs
No ordering	0.4 s	4.3 min	13 min	16 min	15.5 min
One to many to one	0.2 s	3 s	26 s	6.3 min	14.8 min
Random order	0.2 s	3.6 s	18 s	32 s	3.6 min
Bitree order	0.2 s	4 s	1.5 min	7.2 min	16 min

Table 2. Average CPU time of the greedy algorithm

	4 jobs	6 jobs	7 jobs	8 jobs	10 jobs
No ordering	2.5 µs	3.9 µs	4.8 µs	6.4 µs	8.1 µs
One-to-many-to-one	2.5 µs	4.6 µs	5.8 µs	6.9 µs	9.6 µs
Random order	2.6 µs	4.4 µs	6.2 µs	7.7 µs	11.4 µs
Bitree order	2.8 µs	4.5 µs	6 µs	6.9 µs	9.5 µs

for jobs with random partial order, since this type of partial order is usually more constraining than others, and most slowly for the trivial partial order. The greedy algorithm, on the contrary, works faster with trivial partial order, and slower for the random partial order. Still for all types of partial order and for any number of jobs it is much faster than CPLEX.

8   Conclusions

In the paper, the problem of multi-core processor scheduling was analyzed taking into account the bandwidth limitations of the data bus. Two problem formulations are suggested. A mixed integer linear programming model is proposed for the first problem formulation and the MILP solutions were found by CPLEX solver. A greedy algorithm for approximate solving the problem is proposed for the second problem formulation.

The schedules found by the CPLEX package and the greedy algorithm were tested using a program simulating the execution of jobs on the processor cores. The greedy algorithm has only a quadratic running time and a fairly high accuracy: a real-life testing showed that in 83% of the cases the makespan of a greedy schedule deviated from the optimal solution of MILP model less than by 10%, and in 60% of the cases the deviation was less than by 5%. We can conclude that the proposed algorithm of calculation of data bus consumption for a given configuration and the method of calculating the speed of jobs based on these data are close to what happens in real life.

Acknowledgment. The work was funded by project 0314-2019-0019 of Russian Academy of Sciences (the Program of basic research I.5).

References

1. Althaus, E., et al.: Scheduling shared continuous resources on many-cores. J. Sched. **21**(1), 77–92 (2017). https://doi.org/10.1007/s10951-017-0518-0
2. Blazewicz, J., Brauner, N., Finke, G.: Scheduling with discrete resource constraints. In: Leung, J.Y-T. (ed.) Handbook of Scheduling, pp. 23–1–23–18. CRC Press, Boca Raton (2004)
3. Garey, M.R., Johnson, D.S.: Computers and Intractability. A Guide to the Theory of NP-Completeness. W.H. Freeman and Company, San Francisco (1979)

4. Ierapetritou, M.G., Floudas, C.A.: Effective continuous-time formulation for short-term scheduling 1. Multipurpose Batch Processes. Indu. Eng. Chem. Res. **37**(11), 4341–4359 (1998)
5. Intel Hyper-Threading Technology. https://www.intel.com/content/www/us/en/architecture-and-technology/hyper-threading/hyper-threading-technology.html
6. Intel Turbo Boost Technology 2.0. https://www.intel.ru/content/www/us/en/architecture-and-technology/turbo-boost/turbo-boost-technology.html
7. Jiang, Y., Shen, X., Chen, J., Tripathi, R.: Analysis and approximation of optimal co-scheduling on chip multiprocessors. In: Proceedings of the 17th International Conference on Parallel Architectures and Compilation Techniques (PACT 2008), pp. 220–229 (2008)
8. Jozefowska, J., Weglarz, J.: Scheduling with resource constraints - continuous resources. In: Leung, J.Y.-T. (ed.) Handbook of Scheduling, pp. 24–1–24–15. CRC Press, Boca Raton (2004)
9. Liu, M., Chu, F., He, J., Yang, D., Chu, C.: Coke production scheduling problem: a parallel machine scheduling with batch preprocessings and location-dependent processing times. Comput. Oper. Res. **104**, 37–48 (2019)
10. Merkel, A., Stoess, J., Bellosa, F.: Resource-conscious scheduling for energy efficiency on multicore processors. In: Proceedings of the 5th European Conference on Computer Systems (EuroSys 2010), pp. 153–166 (2010)
11. Nesbit, K.J., Aggarwal, N., Laudon, J., Smith, J.E.: Fair queuing memory systems. In: Proceedings of 39th Annual IEEE/ACM International Symposium on Microarchitecture (MICRO 2006), pp. 208–222 (2006)
12. Servakh, V.V.: Effectively solvable case of the production scheduling problem with renewable resources. Diskretn. Anal. Issled. Oper. Ser. **27**(1), 75–82 (2000). (in Russian)
13. Shabtay, D., Kaspi, M.: Parallel machine scheduling with a convex resource consumption function. Eur. J. Oper. Res. **173**(1), 92–107 (2006)
14. Shahrokhi, F., Matula, D.W.: The maximum concurrent flow problem. J. ACM **37**(2), 318–334 (1990)
15. Shioura, A., Shakhlevich, N.V., Strusevich, V.A.: Preemptive models of scheduling with controllable processing times and of scheduling with imprecise computation: a review of solution approaches. Eur. J. Oper. Res. **266**(3), 795–818 (2018)
16. Cirne, W., Desai, N., Frachtenberg, E.,, Schwiegelshohn, U. (eds.): JSSPP 2012. LNCS, vol. 7698. Springer, Heidelberg (2013). https://doi.org/10.1007/978-3-642-35867-8
17. Xiao, Z., Chen, L., Wang, B., Du, J., Li, K.: Novel fairness-aware co-scheduling for shared cache contention game on chip multiprocessors. Inf. Sci. **526**, 68–85 (2020)
18. Zhuravlev, S., Blagodurov, S., Fedorova, A.: Addressing shared resource contention in multicore processors via scheduling. In Proceedings of the 15th International Conference on Architectural Support for Programming Languages and Operating Systems (ASPLOS 2010), pp. 129–142 (2010)

A Novel Algorithm for Construction of the Shortest Path Between a Finite Set of Nonintersecting Contours on the Plane

Alexander Petunin[1,2], Efim Polishchuk[1], and Stanislav Ukolov[1(✉)]

[1] Ural Federal University, Yekaterinburg, Russia
s.s.ukolov@urfu.ru
[2] Institute of Mathematics and Mechanics, UBr RAS, Yekaterinburg, Russia
https://urfu.ru/
https://www.imm.uran.ru/

Abstract. An optimization problem that arises during tool path routing for CNC sheet cutting machine is considered for the case when parts are bounded by line segments and circular arcs and pierce points lay on the bounds. Technique of continuous cutting is used, i.e. each contour is cut as a whole from any starting point. The task of tool path length minimization is reduced to the task of air move length minimization which is shown to be equivalent to finding the shortest broken line with vertices on non-nesting disjoint contours on the plane. The algorithm of building such a broken line for a fixed order of contour processing is devised and proved to deliver local minimum. Some sufficient conditions for this minimum to be global are discussed. A heuristic algorithm for finding the optimal contour cutting order is proposed based on Variable Neighborhood Search approach. Results of a computational experiment and a comparison with the exact solution of GTSP problem are presented.

Keywords: Tool path problem · Continuous cutting problem · Local search · Sufficient conditions of global extremum · Heuristic · Discrete optimization · Variable neighborhood search · GTSP

1 Introduction

A number of optimization problems arise during development of control programs for CNC sheet cutting machines. One of them is the task of minimizing the tool air move, which in some special cases can be reduced to the problem of finding the shortest polyline with vertices on flat contours. Contours are interpreted as the boundaries of flat parts. The location of the contours on the plane is determined during the solution of the "nesting" problem. Both tasks are generally NP-hard.

This work was supported by the Russian Foundation for Basic Research under Grant 20-08-00873.

In its turn, the task of minimizing tool air move is a subtask of another optimization problem – the task of optimizing the tool path when cutting flat parts. Its exact solution cannot be obtained for problems that actually arise in production (for hundreds of parts/contours) in a reasonable time, therefore, various heuristics are typically applied to get solutions of acceptable quality. At the same time, the issues of developing algorithms that provide optimal solutions for some problem cases, as well as evaluating the quality of their solutions in comparison with the optimal solution, remain unresolved and are of significant scientific interest.

The general problem of optimizing the tool path when cutting 2D objects on CNC machines, which consists in minimizing cutting time and cost, includes a whole range of different optimization tasks. A classification of such problems can be found in [9,13,22], see Fig. 1.

- Continuous Cutting Problem (CCP): each closed contour (that bounds a part) is cut out entirely by one movement of the torch, but cutting can start from any point (and finishes at the same point).
- Generalized Traveling Salesman Problem (GTSP): cutting can start only at one of the predefined points on the contour, the contour must be cut entirely.
- Endpoint Cutting Problem (ECP): cutting can start only at one of the predefined points on the contour, and the contour can be cut in several approaches, in parts.
- Segment Continuous Cutting Problem (SCCP): the notion of a cutting segment is introduced, which is a generalization of a contour; it can be either a part of a contour or a combination of several contours or their parts. Each segment is cut out entirely, thus $CCP \subset SCCP$.
- Generalized Segment Continuous Cutting Problem (GSCCP): segment cutting (SCCP), but the selection of segments is not fixed in advance, but is subject to optimization
- Intermittent Cutting Problem (ICP): the most general cutting problem described in the literature, when contours can be cut in parts, in several approaches, and cutting can begin at any point in the contour.

Tool path optimization problems in practice often reduce to discrete optimization problems by discretizing the contours to be cut with a certain step ε, that is, they reduce to ECP [8,14,24] or its special case, GTSP [3,18,27,28]. CCP can also be reduced to GTSP. In this case, however, the total error in the air move length reaches $N \cdot \varepsilon$, where N is the number of contours. To guarantee the accuracy of the result of δ, it is necessary to choose a small $\varepsilon \approx \delta/N$, so the total number of points on the contour grows (as $O(N)$) and the exhaustive search becomes exponential. Nevertheless, such problems can be successfully solved, for example, by the dynamic programming (DP) method, for small $N \approx 30$ even precisely (see, in particular [6]).

Tool path routing without using discretization (CCP) is further considered in this paper. The publications on this subject are rare. [1,26] can be noted, where heuristic algorithms are proposed.

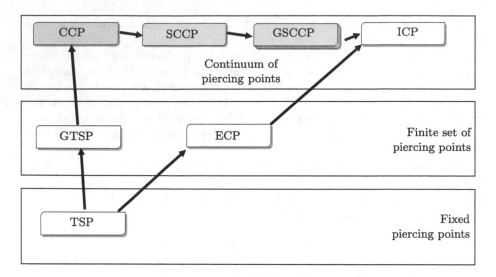

Fig. 1. Classification of Cutting Problems

1.1   Technological Constraints

The need to execute the tool path on a CNC sheet cutting machine imposes a number of technological limitations on it.

The so-called "precedence constraint" is by far most popular in the literature. It is caused by the fact that after cutting a closed contour, its interior is usually not held by anything and can freely shift, rotate and even fall. For this reason, the internal contours of parts must be cut before the external contours containing them, and parts located in the holes of large parts even earlier.

Finally, most cutting technologies require that the cutting not be carried out strictly along the contour, but with some indentation. This shift can be performed both during the solution of the routing problem, and after – at the stage of generating the control program for the CNC cutting machine or even by the machine itself during the cutting process. In addition, the pierce point (tool switch-on point) should generally be located even further from the contour to avoid part damage. However, this work completely ignores this requirement. Thus, it is further assumed that the tool moves exactly along the contour of the part and the pierce point is located directly on the contour (as well as the switch-off point of the tool).

2   Continuous Cutting Problem

Consider the Euclidean plane $\mathbb{R}^2$ and its region B bounded by a closed contour (rectangle in most cases), which is a model of the sheet material to be cut. Let N pairwise disjoint flat contours $\{C_1, C_2, ..., C_N\}$ be given inside B, bounding

n parts $\{A_1, A_2, ..., A_n\}$. A part can be limited by either one contour or several (external and internal holes), so that in general $n \le N$.

The contours C_i can have an arbitrary shape, but we will only consider the case when they consist of (a finite number of) segments of lines and arcs of circles, which is determined by the existing technological equipment. In case when the contours consist only of line segments, the continuous cutting problem is reduced to one of the variants of the Touring Polygon Problem (TPP), see [10].

Further, two points are set in region B (usually at its boundary), we denote them as M_0, M_{N+1} (almost always $M_0 = M_{N+1}$), which represent the beginning and end of the cutting route.

Continuous Cutting Problem is to find:

1. N pierce points $M_i \in C_i, i \in \overline{1, N}$
2. Contour C_i traversal order, i.e. permutations of N elements $I = (i_1, i_2, ..., i_N)$

The result of solving the problem will be the route $\{M_0, M_{i_1}, M_{i_2}, ... M_{i_N}, M_{N+1}\}$. The objective function in this case is greatly simplified in comparison with the general cutting problem and is reduced to minimizing the air move length.

$$\mathcal{L} = \sum_{j=0}^{N} |M_{i_j} M_{i_{j+1}}| \tag{1}$$

$$\mathcal{L} \to \min$$

Where, for sake of simplicity, we introduce the notation $M_{i_0} = M_0$, $M_{i_{N+1}} = M_{N+1}$.

In addition, we will solve the optimization problem with an additional constraint, the so-called "precedence constraint". Although the contours C_i do not intersect, they can be nested into each other, i.e., $\tilde{C}_a \subset \tilde{C}_b$, where $\tilde{C}_a$ denotes a 2-dimensional figure bounded by the contour C_a (in the more familiar notation $C_a = \partial \tilde{C}_a$). In the general tool path routing problem, this can be caused by two different circumstances (holes in parts and placement of smaller parts in holes larger to save material), but in this case these options are processed the same way.

If one contour is located inside another, then the nested contour must be cut out (visited) earlier than the outer one: $\tilde{C}_a \subset \tilde{C}_b \Rightarrow i_a < i_b$ in the permutation $I = (i_1, i_2, ..., i_N)$. Thus, not all permutation of the contours are feasible.

3   CCP-Relax Algorithm to Solve Continuous Cutting Problem

The proposed solution algorithm consists of several stages, easily associated with the nature of the problem being solved:

1. **Removal of external contours.** To automatically comply with the precedence constraint, we start by removing all contours containing nested ones. This generally leads to a reduction (significant in some cases) of the size of the problem (from N to some N'), and thus reduces the calculation time in the second and especially the third stage.

2. **Continuous optimization.** Assuming the order of contours processing $I = (i_1, i_2, ..., i_N)$ fixed we look for the coordinates of the pierce points $M_i \in C_i$, minimizing the total air move length (1).
   For every pierce point M_i we find it's optimal position, while others remain motionless. This relaxation is repeated a few times until converged. In practice, it happens very fast in $O(N)$ time and is therefore used as a subroutine in the next step.

3. **Discrete optimization.** We use Variable Neighborhood Search (VNS, see [12]) to find contours processing order $I = (i_1, i_2, ..., i_N)$.
   This step in fact solves famous *Travelling Salesman Problem* with special distance function, calculated at the previous step:

$$\mathcal{L}(I') = \min_{M_1, M_2 ... M_N} \mathcal{L}(M_1, M_2 ... M_N | I')$$

Note, that other heuristics for discrete optimization may be used at this step as well. For instance, one can use modern solvers to first solve GTSP problem, associated with CCP, and then apply continuous relaxation (previous step) to convert solution of GTSP to that of CCP. This idea deserves further investigation.

4. **Recovery of removed contours.** Having got the tool path that visits "inner" contours (remained after first step), we find piercing points for other contours by simple intersecting them with the tool path. Of multiple points we select one (for each contour) so as to meet precedence constraint.
   This is straightforward step of linear time complexity.

For detailed explanation of the CCP-Relax algorithm steps refer to [20].

3.1   Optimality of Continuous Optimization Problem Solution

From a practical point of view, the described algorithm turns out to be quite workable – it generates high-quality tool path routes in an acceptable time, but this is an empirical result. The theoretical justification of the properties of the resulting routes is interesting. The greatest difficulty is, of course, the third step of the algorithm – discrete optimization, both from a theoretical and a practical point of view. This work focuses the second step of the algorithm – continuous optimization.

Remark 1. Figure 2 shows an example where a trajectory that is not improved by shifts of vertices individually may not deliver a global minimum.

We were able to formulate some statements regarding the quality of continuous optimization solutions at Step 2 of CCP-Relax algorithm. We present them here without proof, which will be published in a separate paper.

Fig. 2. Two tool paths delivering local minimum

We consider the case of fixed order of contours processing $I = (i_1, i_2, ..., i_N)$. and apply Step 2 of CCP-Relax algorithm to get broken line L_*, visiting all the contours C_i in the said order.

Proposition 1. *If we move several adjacent vertices of the broken line L_* so that they remain on the same segments of the contours, then the length of the resulting broken line will not decrease.*

This statement means that the algorithm always delivers a local minimum, however not yet global, as for example in Fig. 2.

To guarantee the latter, the following sufficient condition may be required:

Condition 1. *Let **one** of the following requirements be satisfied for every piercing point M_i:*

1. *Segment $M_{i-1}M_{i+1}$ intersects the contour C_i, i.e. $M_i \in M_{i-1}M_{i+1}$*
2. *The tangent at M_i to the ellipse with foci M_{i-1} and M_{i+1} and passing through M_i separates the ellipse and the contour C_i.*

Proposition 2. *Let Condition 1 is satisfied for (every vertex of) L_*.*

If we move several adjacent vertices of the broken line L_* so that they remain on the contours, then the length of the resulting broken line does not decrease, that is, the broken line L_* delivers a global minimum.

Remark 2. Suppose that besides the trajectory L_*, there is another trajectory delivering a global minimum. Then it follows from the last statement that they coincide as lines, that is, the difference can only be at the points of intersection with the contours.

Condition 1 is easily verified programmatically, but it can be simplified so that in most practical cases to be checked simply visually.

Condition 2. *When segment $M_{i-1}M_{i+1}$ doesn't intersects the contour C_i but*

1. *If the vertex M_i is the internal point of the linear segment of the contour and the entire contour C_i is on one side of the that segment line (which is the tangent from Condition 1; otherwise there must be a better $M_i' \in C_i$).*
2. *If the vertex M_i is terminal (belongs to two linear segments of the contour; is also vertex of C_i), and the entire contour is inside the corner with the rays from the point M_i along these segments.*
3. *If the region $\tilde{C}_i$ bounded by the contour C_i is convex.*

4   New Approach to Intermittent Cutting Problem

Intermittent Cutting Problem is the most complex and general of all varieties of cutting problems. It can be approached to both from theoretical positions as well as by using some practical techniques.

In addition to standard cutting technique (which in fact leads to Continuous Cutting Problem), some others are often used, for instance, "multi-segment" and "multi-contour" cutting. The former cuts single contour of a part in several passes, using several piercing points. The latter cuts a few contours at once, as seen at Fig. 3.

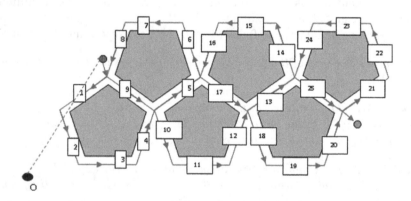

Fig. 3. Example of complex cutting segment for six parts/contours

In order to apply those technique we introduce a notion (see [21]):

Cutting Segment $S = \overrightarrow{MM^*}$ is a tool trajectory from piercing point M up to point of switching tool off M^*.

Cutting segment is used to contain single contour, but this is not the case any more. It also can be a part of contour (for multi-segment cutting) as well as several contours at once (i.e. multi-contour cutting).

In fact, multi-contour cutting example at Fig. 3 can also represent a single cutting segment in some bigger cutting problem instance.

Since the cutting direction is defined for the cutting segment, we need a more general concept:

Basic Segment B^S is a part of cutting segment $S = \overrightarrow{MM^*}$ without lead-in and lead-out trajectory (the very beginning and ending parts of segment, where tool approaches contour of a part and leaves it). Basic segment has no direction and contains only geometry of contours to cut.

Using the concept of basic segment, we can now formulate a generalization of CCP:

Segment Continuous Cutting Problem ($SCCP$) is a cutting problem with fixed set (as well as number of) basic cutting segments: $SCCP = \{B^{S_i}\}$.

CCP-Relax algorithm described above can be applied to solve SCCP problem in the same way as for CCP problem for which it was originally designed.

And now, note that for predefined nesting (i.e. fixed positioning of parts' contours on the plain), the whole *ensemble* of basic segments can be generated by combining and dividing contours into different segments. See, for instance, Fig. 4, where multi-contour segments are filled with black color. This leads us to even more general:

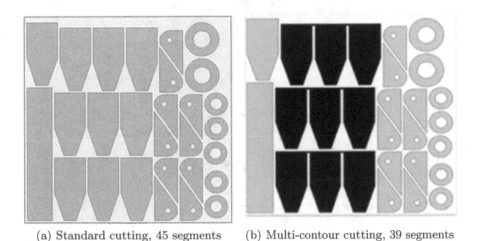

(a) Standard cutting, 45 segments   (b) Multi-contour cutting, 39 segments

Fig. 4. *Ensemble* of Segment Cutting Problems

Generalized Segment Continuous Cutting Problem ($GSCCP$) is that *ensemble* of several $SCCP$ problems for the same nesting: $GSCCP = \{SCCP_i\}$.

By introducing the class of $GSCCP$, we have significantly expanded the existing classification of tool path problem for CNC sheet cutting machines. Actually $SCCP$ and $GSCCP$ are ICP subclasses containing all tasks with finite sets of basic cutting segments, i.e. $CCP \subset SCCP \subset GSCCP \subset ICP$.

General Scheme for GSCCP Solving

Assuming an ensemble $\{SCCP_i\}$ of base segment sets $SCCP_i = \{B^{S_j}\}$, $i \in \overline{1,T}, j \in \overline{1,K_i}$ to be known, the following scheme for GSCCP solving is presented:

- Each task $SCCP_i$ is solved independently with one of existing algorithms, for instance:
  1. *CCP-Relax*, heuristic described above in Sect. 3.
  2. *DP-GTSP*, exact algorithm based on Dymaic programming for the case of relatively small problem dimensions, see [6]
  3. *Greedy-GTSP*, iterative greedy heuristic algorithm, see [19]
  For discrete algorithm to use, cutting segments can be pre-sampled as shown at Fig. 5.
- The best solution is selected according to objective function (1).

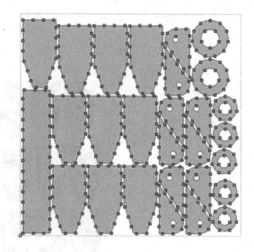

Fig. 5. Corresponding GTSP problem for (S)CCP problem of Fig. 4, 425 points

For example, Fig. 6 shows two solutions of $SCCP$ problems from Fig. 4 given by *CCP-Relax* algorithm. It is easy to see that the two routes are indeed different. Furthermore, the difference can be even more significant in a practical sense due to different numbers of piercing points, since that operation is rather expensive both in terms of time and cost.

5   Numerical Experiments

The quality assessment of the solutions of the described algorithm was carried out on several cutting plans containing real parts. As a comparison base, we used DP algorithm (see [6]) for solving the GTSP problem, which gives an exact solution for small number of contours and special version of GNLS heuristic [25].

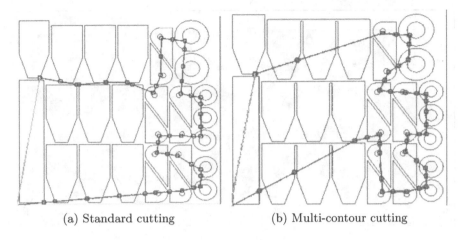

(a) Standard cutting   (b) Multi-contour cutting

Fig. 6. Solution of GSCCP Problem at Fig. 4

Figure 7 shows the exact solution, possible positions of the pierce points are visible. Figure 8 shows the solution to the CCP problem for the same cutting plan.

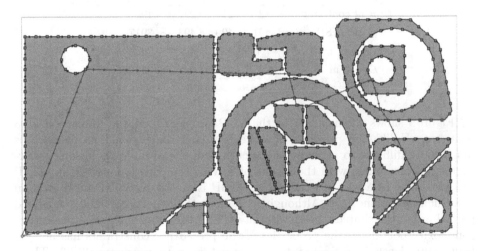

Fig. 7. Exact solution of GTSP, Job #464

It can be seen that both algorithms generated almost identical routes. The main difference is caused by the discretization process to obtain the GTSP task. Because of this, the segments of the route that are straight in the CCP solution turn out to be slightly broken in the GTSP solution, hence total air move length is slightly larger. Numerically, this is shown in Table 1 for several cutting plans.

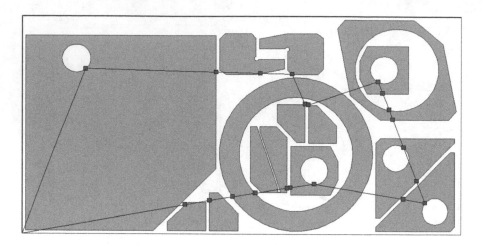

Fig. 8. Solution of CCP, Job #464

Table 1. Solution quality comparison

Job	#229	#464	#3211	#20205
# of parts	11	14	17	115
# of contours	12	21	22	198
# of GTSP points	491	429	493	3917
$\mathcal{L}_{GTSP}$, m	7.729	4.743	4.557	26.098
$\mathcal{L}_{CCP}$, m	7.727	4.706	4.536	25.987

Figure 9 shows the solution to the CCP problem for large dimension (198 contours). Unlike the previous example, for large-dimensional problems it is much more difficult to evaluate the accuracy of the obtained solution. Nevertheless, a comparison with the results of solving the corresponding task GTSP can also serve as a way of estimation. GTSP is known to be NP-hard even on the Euclidean plane [17]. Although it is clear that the bigger the predefined partial order, the simpler the appropriate GTSP task, dependence of theoretical complexity bounds on the properties of the precedence constraints has not yet been insufficiently investigated. In this regard, we note two papers [7,23]. There are two special types of the precedence constraints, for which polynomial time complexity of the GTSP is proven theoretically. The first type of constraints was introduced by E. Balas [2] for the classic TSP. Efficient exact algorithms for the GTSP with precedence constraints of this type are proposed in recent papers [4,5]. Tours that fulfill constraints of the second type are referred to as quasi- and pseudo-pyramidal. Efficient parameterized algorithms for the GTSP with such precedence constraints are proposed in [15,16]. In view of the above, we can summarize that in the field of algorithmic analysis, the GTSP still remains weakly explored. In particular, the absence of efficient Mixed Integer Linear Program (MILP) models for the GTSP makes it impossible to use modern optimizers like Gurobi [11] for construction lower and upper bounds and examining the heuristic solutions. This issue is also pending.

Fig. 9. Example of large problem solution, Job #20205

6 Conclusion

1. The problem of minimizing tool air move of CNC sheet cutting machines for the routing problem from the CCP class is shown to be reduced to a problem without precedence constraint, which reduces the number of contours and the operating time of the algorithm
2. A heuristic algorithm for solving the CCP problem is proposed that does not use contour discretization.
3. It was proved that the CCP-Relax algorithm for finding piercing points for a fixed order of traversing the contours delivers a local minimum.
4. Several easily verified sufficient conditions are formulated and proved for the specified algorithm to deliver global minimum of air move length.
5. CCP-Relax algorithm can be applied to solving more general problems SCCP (Segment Cutting) and GSCCP (Generalized Segment Cutting), thus approaches to solving general ICP cutting problem can be developed on its basis.

The direction of further research is the development of the algorithm for the general case where the pierce points are located outside the contours according to the technological requirements of sheet cutting.

References

1. Arkin, E.M., Hassin, R.: Approximation algorithms for the geometric covering salesman problem. Discret. Appl. Math. **55**(3), 197–218 (1994). https://doi.org/10.1016/0166-218X(94)90008-6

2. Balas, E.: New classes of efficiently solvable generalized traveling salesman problems. Ann. Oper. Res. **86**, 529–558 (1999). https://doi.org/10.1023/A:1018939709890
3. Chentsov, A.G., Chentsov, A.A.: A discrete–continuous routing problem with precedence constraints. Proc. Steklov Inst. Math. **300**(1), 56–71 (2018). https://doi.org/10.1134/S0081543818020074
4. Chentsov, A.G., Khachai, M.Y., Khachai, D.M.: An exact algorithm with linear complexity for a problem of visiting megalopolises. Proc. Steklov Inst. Math. **295**(1), 38–46 (2016). https://doi.org/10.1134/S0081543816090054
5. Chentsov, A., Khachay, M., Khachay, D.: Linear time algorithm for precedence constrained asymmetric generalized traveling salesman problem. IFAC-PapersOnLine **49**(12), 651–655 (2016). https://doi.org/10.1016/j.ifacol.2016.07.767
6. Chentsov, A.G., Chentsov, P.A., Petunin, A.A., Sesekin, A.N.: Model of megalopolises in the tool path optimisation for CNC plate cutting machines. Int. J. Prod. Res. **56**(14), 4819–4830 (2018). https://doi.org/10.1080/00207543.2017.1421784
7. Chentsov, A.G., Grigoryev, A.M.: A scheme of independent calculations in a precedence constrained routing problem. In: Kochetov, Y., Khachay, M., Beresnev, V., Nurminski, E., Pardalos, P. (eds.) DOOR 2016. LNCS, vol. 9869, pp. 121–135. Springer, Cham (2016). https://doi.org/10.1007/978-3-319-44914-2_10
8. Dewil, R., Vansteenwegen, P., Cattrysse, D.: Construction heuristics for generating tool paths for laser cutters. Int. J. Prod. Res. **52**(20), 5965–5984 (2014)
9. Dewil, R., Vansteenwegen, P., Cattrysse, D.: A review of cutting path algorithms for laser cutters. Int. J. Adv. Manuf. Technol. **87**(5), 1865–1884 (2016). https://doi.org/10.1007/s00170-016-8609-1
10. Dror, M., Efrat, A., Lubiw, A., Mitchell, J.S.: Touring a sequence of polygons. In: Proceedings of the Thirty-Fifth Annual ACM Symposium on Theory of Computing, pp. 473–482. ACM (2003)
11. Gurobi Optimization: Gurobi optimizer reference manual (2020). http://www.gurobi.com
12. Hansen, P., Mladenović, N., Moreno Pérez, J.A.: Variable neighbourhood search: methods and applications. Ann. Oper. Res. **175**(1), 367–407 (2010). https://doi.org/10.1007/s10479-009-0657-6
13. Hoeft, J., Palekar, U.S.: Heuristics for the plate-cutting traveling salesman problem. IIE Trans. **29**(9), 719–731 (1997). https://doi.org/10.1023/A:1018582320737
14. Imahori, S., Kushiya, M., Nakashima, T., Sugihara, K.: Generation of cutter paths for hard material in wire EDM. J. Mater. Process. Technol. **206**(1), 453–461 (2008). https://doi.org/10.1016/j.jmatprotec.2007.12.039
15. Khachai, M.Y., Neznakhina, E.D.: Approximation schemes for the generalized traveling salesman problem. Proc. Steklov Inst. Math. **299**(1), 97–105 (2017). https://doi.org/10.1134/S0081543817090127
16. Khachay, M., Neznakhina, K.: Complexity and approximability of the Euclidean generalized traveling salesman problem in grid clusters. Ann. Math. Artif. Intell. **88**(1), 53–69 (2019). https://doi.org/10.1007/s10472-019-09626-w
17. Papadimitriou, C.H.: Euclidean TSP is NP-complete. Theor. Comput. Sci. **4**, 237–244 (1977)
18. Petunin, A.A., Chentsov, A.A., Chentsov, A.G., Chentsov, P.A.: Elements of dynamic programming in local improvement constructions for heuristic solutions of routing problems with constraints. Autom. Remote Control **78**(4), 666–681 (2017). https://doi.org/10.1134/S0005117917040087

19. Petunin, A.A., Chentsov, A.G., Chentsov, P.A.: About routing in the sheet cutting. Bull. South Ural State Univ. Ser. Math. Model. Program. Comput. Softw. **10**(3), 25–39 (2017). https://doi.org/10.14529/mmp170303

20. Petunin, A.A., Polishchuk, E.G., Ukolov, S.S.: On the new algorithm for solving continuous cutting problem. IFAC-PapersOnLine **52**(13), 2320–2325 (2019). https://doi.org/10.1016/j.ifacol.2019.11.552

21. Petunin, A.: General model of tool path problem for the CNC sheet cutting machines. IFAC-PapersOnLine **52**(13), 2662–2667 (2019)

22. Petunin, A.A., Stylios, C.: Optimization models of tool path problem for CNC sheet metal cutting machines. IFAC-PapersOnLine **49**(12), 23–28 (2016)

23. Saliy, Y.V.: Influence of predestination conditions on the computational complexity of solution of route problems by the dynamic programming method. Bull. Udmurt Univ. Maths. Mech. Comput. Sci. (1), 76–86 (2014)

24. Sherif, S.U., Jawahar, N., Balamurali, M.: Sequential optimization approach for nesting and cutting sequence in laser cutting. J. Manuf. Syst. **33**(4), 624–638 (2014)

25. Smith, S.L., Imeson, F.: GLNS: an effective large neighborhood search heuristic for the generalized traveling salesman problem. Comput. Oper. Res. **87**, 1–19 (2017). https://doi.org/10.1016/j.cor.2017.05.010

26. Vicencio, K., Davis, B., Gentilini, I.: Multi-goal path planning based on the generalized Traveling Salesman Problem with neighborhoods. In: 2014 IEEE/RSJ International Conference on Intelligent Robots and Systems, pp. 2985–2990 (2014). https://doi.org/10.1109/IROS.2014.6942974

27. Ye, J., Chen, Z.G.: An optimized algorithm of numerical cutting-path control in garment manufacturing. Adv. Mater. Res. **796**, 454–457 (2013). https://doi.org/10.4028/www.scientific.net/AMR.796.454

28. Yu, W., Lu, L.: A route planning strategy for the automatic garment cutter based on genetic algorithm. In: 2014 IEEE Congress on Evolutionary Computation (CEC), pp. 379–386 (2014). https://doi.org/10.1109/CEC.2014.6900425

The Polyhedral-Surface Cutting Plane Method of Optimization over a Vertex-Located Set

Oksana Pichugina[1]([⊠]) [iD], Liudmyla Koliechkina[2] [iD],
and Nadezhda Muravyova[3] [iD]

[1] National Aerospace University "Kharkiv Aviation Institute",
17 Chkalova Street, Kharkiv 61070, Ukraine
oksanapichugina1@gmail.com
[2] University of Lodz, Uniwersytecka Str. 3, 90-137 Lodz, Poland
liudmyla.koliechkina@wmii.uni.lodz.pl
[3] South Ural State University, 76 Lenin Prospekt, 454080 Chelyabinsk, Russia
muravevanv@susu.ru

Abstract. The Boolean set, permutation vector's sets and many others belong to a class of vertex-located sets (VLS) as they coincide with a vertex set of their convex hull. A polyhedral-surface cutting plane method (PSCM) for linear constrained optimization over VLS is offered. It utilizes representability of a VLS as an intersection of a strictly convex surface S with a polytope P. PSCM applies iteratively two steps dealing with a polyhedral or a surface relaxation of the original problem. First, a polyhedral relaxation is solved on P, and its solution x is verified on belongingness to S. If it holds, the original problem has been solved. Otherwise, a surface relaxation is considered, and a cut of x is formed utilizing a polyhedral cone with apex at x given by active P-constraints and an intersection of its extreme rays with the circumsurface S. Three versions of PSCM and two ways to form the cuts are presented and illustrated. Applicability of PSCM to solve permutation-based and Boolean linear optimization problems is justified. Area of practical applications of the results is indicated.

Keywords: Linear combinatorial optimization · Cutting plane method · Vertex-located set · Polyhedral relaxation · Surface relaxation · Circumsurface

1 Introduction

Cutting plane methods (CPM) play a special role in Integer Programming and Convex Optimization. Their main advantage is that they iteratively decrease a search domain to a convex hull of feasible region E of an optimization problem until a current point x be a feasible point and, respectively, an optimal solution. In each iteration, CPM does not require evaluation constraints' functions

© Springer Nature Switzerland AG 2020
N. Olenev et al. (Eds.): OPTIMA 2020, CCIS 1340, pp. 84–98, 2020.
https://doi.org/10.1007/978-3-030-65739-0_7

and only query a separating oracle, if a condition $x \in E$ is met, and search for a cutting plane for a point x if the condition does not hold. This makes CPM attractive when dealing with large-dimension problems and problems with numerous constraints. However, the need to solve a linear programming relaxation in polynomial time, querying the oracle, and finding the cut, as well as issues of slow convergence of CPM restrict the area of applications.

Typically, CPM is associated with Integer Linear Programming, where LP relaxation is solved on each iteration, the oracle verifies the integrity of x-coordinates, and a Gomory's cutting plane is added to constraints constructed in such a way that new current point has integral coordinates and the number of the coordinates increases throughout the iterative process. Gomory's cuts use an absence of integral points inside integer-grid cells. If a linear combinatorial problem is solved instead of integer optimization one, where a feasible region E is a finite point configuration [8] associated with a set of combinatorial nature (permutations, partial permutations, combinations, etc.), Gomory's cuts need adaptation, which takes into account an absence of feasible points in some nodes of the integer grid and ensures convergence of the method to $x \in E$. These "combinatorial" modifications of the Gomory Cutting Plane Method require using essentially structural properties if E is highly dependent on a type of combinatorial nature of the set as well as on geometric properties of E and associated combinatorial polytope $P = conv E$. Respectively, they need deep studying structural and geometric properties of images of combinatorial spaces in Euclidean space (further referred to as $\mathcal{C}$-sets [18]), as well as the behavior of various classes of functions on the sets. These two research fields, along with developing combinatorial optimization algorithms and a search for applications are an area of research of Euclidean Combinatorial Optimization (ECO) [24–26].

This paper is dedicated to developing a CPM for linear combinatorial programs on sets inscribed into a convex surface. It generalizes and extends results on CPM for combinatorial optimization [1,5,19,20] and results of our work related to exploring properties of such sets and their applications [15,17,18,21].

2   Problem 1 Statement and Properties

Consider the following discrete optimization problem:

$$\text{minimize } cx, \tag{1}$$

$$\text{subject to } Ax \leq B, \tag{2}$$

$$x \in E \subset \mathbb{R}^n, \tag{3}$$

$$1 < |E| < \infty, \tag{4}$$

where $A \in \mathbb{R}^{m \times n}$, $c, x \in \mathbb{R}^n$, $B \in \mathbb{R}^m$,

$$m \text{ is fixed.} \tag{5}$$

In addition, there exists $f : \mathbb{R}^n \to \mathbb{R}^1$, f is a strictly convex function such that

$$S = \{x \in \mathbb{R}^n : f(x) = 0\}, \tag{6}$$

$$E \subset S. \tag{7}$$

Also, assume that there exists a polynomial separating oracle (further referred to as *the oracle*):

$$\exists \phi(x, P) \tag{8}$$

examining if a point $x \in \mathbb{R}^n$ belongs to a polytope:

$$P = conv\ E \tag{9}$$

(further referred to as a combinatorial polytope). If not, it generates a cutting plane for the point x in the form of a P-facet inequality.

Problem (1)–(4) is a generic linear combinatorial optimization problem (LCOP) on a set E, which is not a singleton. The additional constraints (6), (7) means that E is inscribed into a strictly convex surface [22], i.e., the one given by a strictly convex function $f(x)$. Note that from these two constraints follows that the search domain E coincides with a vertex set of P. Following the terminology introduced in [28], E is a vertex-located set (VLS), i.e. $E = vert\ P$. Moreover, the conditions (6), (7) imply that E is a surface located set (SLS) [17]. Thus, we pose LCOP on an SLS-subclass of VLS satisfying (8). That is why, when solving the problem (1)–(8) (further referred to **Problem 1**) we will use features of the oracle as well as of a circumsurface S and its inducing function $f(x)$.

Problem 1 belongs to a class of ECO-problems [25], for which powerful tools are developed based on combining Euclidean space properties with structural features of special classes of E [13,15–17,24,26,29]. Problem 1 is a LCOP on a VLS E. For its exact solutions, ECO-methods such as the Combinatorial Cutting Plane Method (CCM) [5] and different Branch & Bound (B&B) techniques [11,27] are applicable. The additional assumption about the surfaced locality of E allows using the Polyhedral-Surface Methods (PSM) for optimization, namely, the Greedy PSM for an approximate solution and the Branch and Bound PSM for exact [16,17]. Note that these PSM require adaptation to constrained problems. An important step of the PSM-implementation is solving a polyhedral relaxation problem, where (3) is replaced by $x \in P$ (further **Problem 2**) yielding a linear program on P. In order to solve Problem 2 easier than Problem 1, some properties of P need to be used, such as its H-representation (further referred to as *Property 1*), the separating oracle (8) (further *Property 2*). Note that Property 2 means that Problem 2 is polynomially solvable.

A special mathematical field – Polyhedral Combinatorics – works on deriving H-representations of polytopes associated with combinatorial sets embedded in Euclidean space in order to utilize them in linear optimization [2]. Another direction is a search of the separating oracles of polynomial complexity for combinatorial polytopes (further **Problem 3**) [10,23,26] since it is allows single outing classes of combinatorial problems whose polyhedral relaxation is polynomially solvable by the Ellipsoid Method.

Many combinatorial sets having Properties 1, 2 are known [15,18,21].

Despite the presence of constraints (6)–(8), Problem 1 covers a vast class of practical and theoretical problems including linear Boolean and permutation-based optimization problems [4,7,14,17], which practical applications include telecommunication, VLSI design, warehouse location, military defence, social networks, molecular interaction networks, image processing, computer vision, scientific computing, sparse matrix computation, physics, parallel programming, compiler optimization, load balancing, route planning, and many other problems of optimal planning and geometric design [2,4,6,10,12,24].

Indeed, it is known [16] that the Boolean vector set (the Boolean C-set) $B_n = \{0,1\}^n$ is inscribed in a hypersphere centered at point $(0.5, ..., 0.5)$, namely, $B_n \subset S = S_r(0.5 \cdot \mathbf{e})$, where $\mathbf{e}$ is a vector of units, $r = \frac{\sqrt{n}}{2}$.

At the same time, a set of permutation vectors induced by a multiset G (the general multipermutation C-set [18,21]): $E_{nk}(G) = \{x = (x_1, ..., x_n) : \{x_1, ..., x_n\} = G\}$, where $G \subset \mathbb{R}^1$, $G = \{\{g_i\}_{i \in J_n = \{1,...,n\}} : g_1 \leq ... \leq g_n\}$, a ground set of G is $S(G) = \{\{e_i\}_{i \in J_n} : e_1 < ... < e_k\}$, is inscribed into a family of hyperspheres centered on the ray $a\mathbf{e}$, where $a \in \mathbb{R}^1$ is a parameter, namely, $E_{nk}(G) \subset S = S_{r(a)}(a \cdot \mathbf{e})$, where $r(a) = \left(\sum_{i=1}^n (g_i - a)^2\right)^{1/2}$.

So, in Boolean and permutation-based linear problems, a search domain:

$$E \in \{B_n, E_{nk}(G)\} \tag{10}$$

is SLS called a spherically located set (SpLS) [20]. In addition, for polytopes $D_n = conv\ B_n$, $P_{nk}(G) = conv\ E_{nk}(G)$, the condition (8) is satisfied. Indeed, D_n is the unit hypercube $[0,1]^n$ given by $2n$ constraints, whose feasibility can be easily verified and violated constraints are derived. $P_{nk}(G)$ is a multipermutohedron [21] given by a system $\sum_{i=1}^n x_i = \sum_{i=1}^n g_i, \sum_{i \in \omega} x_i \geq \sum_{i=1}^j g_i, j = |\omega| \subset J_n$ of $2^n - 2$ constraints [29]. Nevertheless, the polytope has Property 2.

Lemma 1. [26] *If $x \in \mathbb{R}^n$, such that*

$$x_1 \leq ... \leq x_n, \tag{11}$$

then $x \in P_{nk}(G)$ if and only if the following constraints are satisfied:

$$\sum_{i=1}^n x_i = \sum_{i=1}^n g_i, \sum_{i=1}^j x_i \geq \sum_{i=1}^j g_i, j \in J_{n-1}. \tag{12}$$

It can be seen that the ordering (11) is polynomially doable as well as verification of the condition (12). Note that (12) consists of $P_{nk}(G)$-constraints only. Thus the oracle (8) is found. If the condition $x \in P$ is violated, it induces a cutting plane for x among constraints (12) of the polytope P.

The subclass (1)–(4), (10) of Problem 1 can be further extended to: sets of permutation and multi-permutation matrices [3]; a signed permutation C-set [15]; an even Boolean C-set [9], Boolean partial permutation C-set [21], and other vertex-located classes of partial permutation C-sets [21]; the even permutation set [29], and other subsets of $E_{nk}(G)$ [21,26].

Note that the conditions (6), (7) of surface locality and (9) of vertex locality are equivalent. However, in an optimization approach described below, a circumsurface S will be taken into account substantially. For the general case, its search is a separate task. If it is possible to find a family of such surfaces, interest is the question of choosing one of them, which utilizing is more beneficial in solving problems on SLS, e.g., Problem 1.

This paper presents a cutting plane method to solve Problem 1, which is based on utilizing the absence of feasible points inside a strictly convex body:

$$C = conv\ S. \tag{13}$$

In particular, from (13), it follows that there are no feasible points in an interior of P and its faces of any dimension.

3 A Polyhedral-Surface Cutting Plane Method Description

Problem 1 is writable in the form of (1), (6)–(8),

$$x \in E', \tag{14}$$

where

$$E' = \{x \in E : Ax \leq B\}, \tag{15}$$

and E satisfies the constraint (4).

Let S' be a convex surface, where E' lies, i.e., there exists a convex function $\varphi(x) : \mathbb{R}^n \to \mathbb{R}^1$ such that

$$S' = \{x \in \mathbb{R}^n : \varphi(x) = 0\}, \tag{16}$$
$$E' \subset S'.$$

Note that such a surface exists, since $E' \subseteq E \subset S'$. As S', either the boundary ΓP of the combinatorial polytope P or the boundary $\Gamma P'$ of the polytope P' can also be chosen, where

$$P' = conv\ E'. \tag{17}$$

Let $\langle x^*, z^* \rangle = \langle x^*, cx^* \rangle$ be an optimal solution to Problem 1.

Remark 1. Without loss of generality, we assume that P' is full-dimensional polytope:

$$\dim P' = n, \tag{18}$$

otherwise, we make a projection into $\dim P'$-space first.

This means that (4) can be replaced by

$$n + 1 < |E| < \infty.$$

Let a convex surface S' is chosen from a set:

$$S' \in \{S, \Gamma P, \Gamma P'\}. \tag{19}$$

3.1 PSCM(S') Outline

- **Step 1.** Initialization: set iteration $j = 0$, $P'^j = P'$.
- **Step 2.** Solve a linear program (1),

$$x \in P'^j, \tag{20}$$

(further referred to as **Problem 2.j**) and, in case of its feasibility, find the problem solution $\langle x^j, z^j \rangle = \langle x^j, cx^j \rangle$. Otherwise, Problem 1 is infeasible. Terminate.

Remark 2. If H-representation of P includes polynomial on n number of constraints, we solve Problem 2.j involving the whole H-representation, otherwise, the consequent inclusion of constraints method (CICM) [26] is applied (see Subsect. 3.2). To the corresponding Problem 1 in these two cases, we will refer to as Problem 1.1 and Problem 1.2, respectively.

- **Step 3.** Check the following condition $x^j \in S$, which is here equivalent to

$$x^j \in E. \tag{21}$$

  - **Step 3.1.** If (21) holds, then $\langle x^*, z^* \rangle = \langle x^j, z^j \rangle$. Respectively, Problem 1 has been solved. Terminate.
  - **Step 3.2.** If (21) does not hold, we construct a cutting plane

$$a^j x \leq b^j \tag{22}$$

  for the point $x^j \notin E$, according to a chosen cutting plane scheme from those described in Subsect. 3.3.
  - **Step 3.3.** Set $P'^{j+1} = \{x \in P'^j : a^j x \leq b^j\}$, $j = j + 1$, go to Step 2.

3.2 CICM for Problem 2.j Solution

We describe how CICM can be applied for solving Problem 2.j and getting a tuple $\langle x^0, z^0 \rangle$.

- **Step 0.** Initialization: set iteration $t = 0$, a search domain is $D'^t = \{x \in D : x \text{ satisfies } (2)\}$, where $D = \prod_{i=1}^{n} [\alpha_i, \beta_i], -\infty < \alpha_i < \beta_i < \infty, i \in J_n$: $D \supset E$.
- **Step 1.** Main stage: solve a linear program (1),

$$x \in D^t. \tag{23}$$

  - If it is infeasible, then Problem 2 is infeasible. Terminate.
  - Otherwise, to the problem solution y^t, apply the oracle $\phi(y^t, P)$.
    . If $y^t \in P$, then $x^j = y^t$, $z^j = cy^t$, hence Problem 2 has been solved. Terminate.
    . If $y^t \notin P$, the violated facet constraint $\bar{a}x \leq \bar{b}$ of P generated by the oracle is added to the current system of constraint yielding:

$$D'^{t+1} = \{x \in D'^t : \bar{a}x \leq \bar{b}\}. \tag{24}$$

  Set $t = t + 1$, go to Step 1.

3.3 Cutting Plane Construction Schemes

In order to form a cutting plane (22), we single out a set:

$$A'^j x \leq B'^j \tag{25}$$

of active P'^j-constraints at point x^j. Normally, (25) includes some additional constraints (2) and a part of P-constraints.

Remark 3. If we deal with Problem 1.2 and apply the CICM for getting x^j, special techniques are needed for deriving the linear system (25), since, generally, a few P-constraints are involved in search of x^j. To the problem of forming this system, we will refer to as Problem 4.

Next, we consider the polyhedral cone (25) (further $Cone_j$) and a part of surface S' cut out by this cone denoting it by S^j: $S^j = S' \cap Cone_j$.

Clearly that $E' \subset S^j$, and the point x^j can be cut off in any way not affecting S^j. At the same time, forming a cut of x^j, we aim to build a strong one. Thus, the cutting plane (22) will be constructed in such a way that

$$a^j x^j > b^j, \tag{26}$$

$$\forall x \in S^j \; a^j x \leq b^j. \tag{27}$$

$Cone_j$ has apex at x^j and edge set:

$$\mathcal{E}_j = \left\{ [x^j, y^{ji}], i \in J_{k_j} \right\}, \tag{28}$$

where

$$N_{S^j}(x^j) = \{y^{ji}\}_{i \in J_{k_j}} \subset S^j \tag{29}$$

is a neighborhood of x^j, $k_j = |N_{S^j}(x^j)|$. Thus, the neighborhood $N_{S^j}(x^j)$ of the point x^j is defined as a set of endpoints of edges of P^j with the origin at x^j and endpoints on the surface S^j. In other words, extreme rays of $Cone_j$ intersect S at points of the set (29).

Assume that a^j, b^j satisfy (26). By construction,

$$N_{S^j}(x^j) \subseteq S^j, \tag{30}$$

hence (27) entails

$$\forall x \in N_{S^j}(x^j) \quad a^j x \leq b^j. \tag{31}$$

From convexity of the surface S^j follows that the reverse is also true, i.e., (31) results in (27). Therefore, further, we replace the conditions (26), (27) by (26), (31) for verification that a cut under consideration is strong and valid.

Lets analyze ways to construct the cut (22) depending on k_j, which satisfies inequality $k_j \geq n + 1$ by (18):

1. If x^j is a nondegenerate vertex of P^j, i.e.,

$$k_j = n + 1, \tag{32}$$

then the bounding hyperplane

$$\Pi^j : a^j x = b^j \tag{33}$$

for the strong valid cutting plane (22), (26), the plane bounding (31) is built through all points (29). In this case, inequality (31) holds as equality:

$$\forall x \in N_{S^j}(x^j) \quad a^j x = b^j. \tag{34}$$

2. If x^j is a degenerate vertex of P^j, i.e.,

$$k_j > n, \tag{35}$$

then there is an ambiguity in the choice of an n-subset $X^j \subset N_{S^j}(x^j)$ through which the hyperplane Π^j is constructed. Following a certain scheme of choosing X^j and constructing a cutting plane presented in Subsect. 3.4, select $X^j : X^j \subset N_{S^j}(x^j) : |X^j| = n + 1$, where X^j are the first best ones with respect to a criterion K.

Now, we form a cutting plane (22) such that the corresponding separating hyperplane (33) passes through the points of X^j:

$$X^j \subset \Pi^j. \tag{36}$$

In this case, (34) is replaced by

$$\forall x \in X^j \quad a^j x = b^j, \tag{37}$$

while the condition

$$\forall x \in N'_{S^j}(x^j) = N_{S^j}(x^j) \backslash X^j \quad a^j x \leq b^j \tag{38}$$

requires verification.

If condition (38) of the validity of the cut (22) does not hold, there exists $y^{ji} \in N'_{S^j}(x^j) : a^j x > b^j$. We replace the worst point $y^{ji'} \in X^j$ according to the criterion K by y^{ji}: $X^j \leftarrow \{y^{ji}\} \cup X^j \backslash \{y^{ji'}\}$.

Now, the process of formation Π^j and validation that the corresponding cutting plane is valid is repeated iteratively until a valid cut (22) is formed.

3.4   Selecting X^j

We will offer two ways to choose the set (36) resulting in forming two types of the cutting plane (22). For that, we introduce ordering of elements of (29) in such a way that

$$z^{j1} \leq z^{j2} \leq \ldots \leq z^{jk_j}, \text{ where } z^{ji} = cy^{ji}, \, i \in J_{k_j};$$

$$r^{ji_1} \leq r^{ji_2} \leq \ldots \leq r^{jik_j},$$

where $z^{ji} = cy^{ji}$, $r^{ji_l} = |x^j - y^{ji_l}|$, $i, l \in J_{k_j}$.

$$\text{Way 1:} \qquad X^j = \left\{x^{ji}\right\}_{i \in J_{n+1}},$$

i.e., X^j includes $n+1$ points of $N_{S^j}(x^j)$ with the smallest values of the objective function. This means that the criterion $K : cx \to \min$.

$$\text{Way 2:} \qquad X^j = \left\{x^{ji_l}\right\}_{l \in J_{n+1}},$$

i.e., X^j consists of $n+1$ of the $N_{S^j}(x^j)$-points closest to x^j. Here, $K : r^{ji} \to \min$.

4   PSCM(S')-Versions

Depending on a choice of the circumsurface S' in (19), we come to three versions of our method.

4.1   Modified Combinatorial Cutting Plane Method (MCCM)

If $S' = \Gamma P'$, we get PSCM($\Gamma P'$). On initial iteration, $j = 0$, the edges (28) go from x^j to adjacent vertices of the polytope P'. Thus the neighborhood (29) of x^0 is a set of all adjacent vertices of P'. To the PSCM($\Gamma P'$), we will refer to as the Modified Combinatorial Cutting Plane Method (MCCM).

Let N be the maximum number of PSCM-iterations. In the title, the following fact is reflected. If we deal with Problem 1.1 and, for all

$$x^j, j \in J_N^0 = J_n \cup \{0\}, \tag{39}$$

(32) holds, i.e., all solutions of polyhedral relaxations of Problem 1 are nondegenerate, our method is reduced to the Combinatorial Cutting Plane Method (CCM) of linear optimization over vertex located sets [5].

In [5], the cutting plane is derived from the last simplex tableau in the form of inequality:

$$\sum_{j \in J} \frac{x_j}{\Theta_j} \geq 1, \tag{40}$$

where J is a set of non-basic variables at x^j, $\Theta_j = \min\limits_{i:a_{ij}>0} \frac{b_i}{a_{ij}}$, $j \in J$.

The disadvantage of CCM is that, in case if among (39) are degenerate vertices, the authors propose to use standard techniques of resolving degeneracy in linear programming such as perturbation. An issue is that it results in replacing the search domain P^j by a simple polytope in the vicinity of x^j, which leads to slow convergence of the method. Note that, in combinatorial optimization, degeneracy is quite common since, among combinatorial polytopes, simple polytopes form a minor subclass. With this regard, MCCM offers a new technique to resolve degeneracy constructing the set $N_{S^j}(x^j)$ and constructing the cut through its proper subset. Another issue preventing utilizing the cutting plane (40) is that it does not analyze all active constraints as x^j in case if Problem 1.2 is solved. In contrast, MCCM assumes that Problem 3 of deriving active constraints (25) has been solved. Let us denote

$$d^{ji} = y^{ji} - x^j, i \in J_{k_j} - \tag{41}$$

direction vectors of the edges (28), where $N_{S^j}(x^j)$ is associated with MCCM.

4.2   Combinatorial Polytope Cutting Plane Method (CPCM)

MCCM utilizes an absence of feasible points in an interior of the polytope P' and its faces. The next PSCM-version uses an observation that no feasible points are in an interior of the combinatorial polytope P as well as inside its faces. Thus choosing $S' = \Gamma P$, we come to PSCM(ΓP). Similarly to MCCM, if $j = 0$, the edges (28) start at x^j and extend to the intersection with a boundary of a polyhedron found from P by excluding its constraints active at x^0. As a result, it yields a set of intersections of extreme rays of $Cone^j$ with a boundary of the combinatorial polytope P as a neighborhood of x^0. Therefore, to this version of PSCM we will refer to as the Combinatorial Polytope Cutting Plane Method (CPCM).

In order to apply CPCM, a problem of finding the closest facet of a combinatorial polytope P to a point x in a certain direction d (further Problem 5) needs to be solved. It can be stated as follows: find a hyperplane $p : \alpha x = \beta$ such as minimize $|y - x|$, where p is a facet of P, $y \in p$, $\exists \lambda > 0 : y = x + \lambda d$.

In order to form $N_{S^j}(x^j) \subset \Gamma P$, Problem 5 is solved for $x = x^j$, $d = d^{ji}$, where d^{ji} is found by (41) for each $i \in J_{k_j}$, $j \in J_N^0$.

4.3   Surface Cutting Plane Method (SCM)

Finally, observe that no point of E in an interior of the convex body (13) and apply it in the final version of PSCM, where $S' = S$. This version PSCM(S) is titled the Surface Cutting Plane Method (SCM). Here, the edges $[x^j, y^{ji}] = = [x^j, x^j + d^{ji}], i \in J_{k_j}$ are extended until extreme rays of $Cone^j$ intersect ΓP first and then reach the surface S. SCM requires solving Problem 6 consisting in finding an intersection of a half-line starting at x^j having the direction vectors (41) with the strictly convex surface S.

5   PSCM Illustration and PSCM-Versions Comparison

In Figs. 1, 2 and 3, the above versions of PSCM are illustrated. Here, the combinatorial polytope P is a reqular pentagon (a dotted line), S is a circle circumscribed around P, there are four additional constraints (2) resulting in forming a heptagon as the polytope P' (shadowed region).

In Fig. 1, an illustration for MCCM is given. It is seen that seven iterations are required to get an optimal solution x^*. Thus, all vertices of P' are explored and utilized. In Fig. 2, CPCM is illustrated. It can be seen that for getting x^* six iterations are sufficient. Finally, in Fig. 3, the work of SCM is shown, in particular, that x^* is found in five iterations.

For x^0, it can be observed that the cutting plane produced by SCM dominates the CMCM-cut. Respectively, the later dominates the cutting plane induced by MCCM. The reason is an extend of prolongation of P'-edges – the highest corresponds to SCM, the next highest one – to CPCM.

The question arises, is it always SCM-cuts dominate CPCM-ones and MCCM-cutting planes are dominated by CPCM-ones.

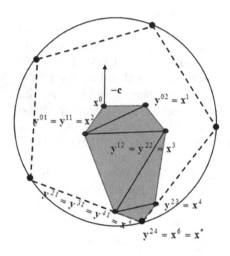

Fig. 1. PSCM($\Gamma P'$) = MCCM   **Fig. 2.** PSCM(ΓP) = CPCM

Theorem 1. *If the polyhedral cone* (25) *is simplicial, i.e.,* (32) *holds, and S, S' are convex surfaces circumscribed around E such that*

$$conv(S) \subset conv(S'),$$

then the PSCM(S')-cut of x^j dominates the PSCM(S)-one.

Proof. Let the neighborhoods of x^j be

$$N_S(x^j) = \{y^{ji}\}_{i \in J_{n+1}}, \; N_{S'}(x^j) = \{y'^{ji}\}_{i \in J_{n+1}}.$$

$cx = cx^j$ is a supporting plane hence $\forall i \in J_{n+1} \; cy^{ji} \geq cx^j$ or, in terms of (41),

$$cd^{ji} \geq 0, i \in J_{n+1}. \tag{42}$$

The points of $N_{S'}(x^j)$ can be represented as follows:

$$y'^{ji} = x + k^{ji}d^{ji}, \text{ where } k^{ji} \geq 1, i \in J_{n+1}. \tag{43}$$

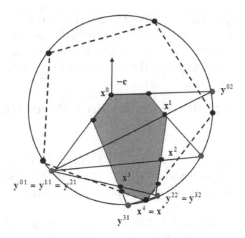

Fig. 3. PSCM(S) = SCM

From that, by (42), (43), $c(y'^{ji} - y^{ji}) = c(x + k^{ji}d^{ji}) - c(x + k^{ji}d^{ji}) = (k^{ji} - 1)cd^{ji} \geq 0, i \in J_{n+1}$, hence

$$cy'^{ji} \geq cy^{ji}, \; i \in J_{n+1}. \tag{44}$$

On the next iteration, x^{j+1} will be an element of $N_{\mathcal{S}'}(x^j)$ or $N_{\mathcal{S}'}(x^j)$, respectively, with the smallest objective function value. In particular, $z^{j+1} = \min_{i \in J_{n+1}} cy^{ji}$, $x^{j+1} = \arg\min_{i \in J_{n+1}} cy^{ji}$ will be a solution to Problem 2.j by PSCM($\mathcal{S}$), while $z'^{j+1} = \min_{i \in J_{n+1}} cy'^{ji}$, $x'^{j+1} = \arg\min_{i \in J_{n+1}} cy'^{ji}$ will be the one by PSCM($\mathcal{S}'$).

From (44) follows that $\min_{i \in J_{n+1}} cy'^{ji} \geq \min_{i \in J_{n+1}} cy^{ji}$, $i \in J_{n+1}$, i.e.,

$$z'^{j+1} \geq z^{j+1}. \tag{45}$$

Thus the cut induced by PSCM($\mathcal{S}'$) dominates the PSCM($\mathcal{S}$)-cutting plane.

Corollary 1. *If the polyhedral cone (25) is simplicial, then the PSCM(ΓP)-cutting plane of x^j dominates the PSCM($\Gamma P'$)-one and is dominated by the PSCM(S)-cutting plane.*

Theorem 1 formulates a sufficient condition of domination SCM-cuts over CPCM-ones, and domination of the later over MCCM-cutting planes. An issue of extending it onto degenerate case is that, in this situation, when the bounding plane (33) intersects extreme rays, the only $n+1$ of them is guaranteed to be on S^j. Because now $z^{j+1} = \min_{i \in J_{k_j}} cy^{ji}$, $z'^{j+1} = \min_{i \in J_{k_j}} cy'^{ji}$, the condition (45) requires additional verification. The advantage of using the SCM-cutting plains does not limited by their depth. Another plus is a possibility of constructing the cutting planes using only a part of active constraints at x^j sufficient for a formation this vertex. The information can be extracted from the last simplex tableau, and solving Problem 4 does not need. It is especially useful when we deal with Problem 1.2, i.e., polytopes under consideration are given by an exponential number of constraints, and we solve the corresponding Problem 2 by CICM.

6 PSCM-Specifics for Boolean and Permutation-Based LCOP

On $E = B_n$, Problem 1 becomes Problem 1.1. Thus Problem 3 is directly solvable, while solutions to Problems 5, 6 can be found in [19]. For the set, CPC and SCM are closely connected to spherical and intersection cuts [1] assuming that this problem is solved as ILP.

On $E = E_{nk}(G)$, Problem 1 belongs to the Problem 1.2-subclass. Solving Problem 2.j, it is suggested to choose $D = \{x \in \mathbb{R}^n : g_1 \leq x_i \leq g_n, i \in J_n, x\mathbf{e} = g_1 + \dots + g_n\}$. Solutions to Problem 4 are directly derived from the last simplex tableau if (11) satisfies $x_1 < \dots < x_n$. If there are repetitions of x-coordinates, some inequalities from other units of $P_{nk}(G)$-inequalities needs verification. A method to solve Problem 5 is presented in [27]. $dim(P_{nk}(G)) = = n - 1$, thus a projection onto $\mathbb{R}^{n-1}$ is needed before applying PSCM. It can be done in two ways: a) an orthogonal projection when the feasible set remains SpLS [24] while a structure of constraints become more complicated; b) a projection resulting in a $n - 1$-partial permutation $\mathcal{C}$-set, which is well studied [29], ellipsoidally located and SpLS for $k = 2$ only. This means that Problem 6 is easly solvable in both the cases.

7 Conclusion

The paper presents the Polyhedral-Surface cutting plane method (PSCM) for linear constrained optimization over a combinatorial set E inscribed into a strictly convex surface (Problem 1). It essentially uses geometric features of a feasible domain, the corresponding polytope P, and the circumsurface S. Among the properties is a vertex locality of E resulting in its representation $E = P \cap S$ and ability to combine in PSCM polyhedral and surface relaxations Problem 1. There are offered two ways of constructing the cutting planes and three variants of PSCM depending on a choice of a surface involved from a set $\{S, \Gamma P, \Gamma P'\}$, where P' is a convex hull of the feasible domain. Two of them generalize well-known spherical and intersection cuts of E. Balas. For a cut of nondegenerate vertex, it is justified domination of PSCM(S)-cuts over the ones induced by PSCM($\Gamma P'$) and PSCM(ΓP). For Boolean and permutation-based problems, the applicability of PSCM is established. Auxiliary problems are formulated required solution for extending PSCM onto other classes of surface located sets. Graphic Illustration of the PSCM-versions is provided, and an area of PSCM-applications is outlined.

References

1. Balas, E.: Intersection cuts: a new type of cutting planes for integer programming. Oper. Res. **19**(1), 19–39 (1971). https://doi.org/10.1287/opre.19.1.19
2. Balinski, M.L., Hoffman, A.J. (eds.): Polyhedral Combinatorics: Dedicated to the Memory of D.R. Fulkerson. Elsevier Science Ltd., Amsterdam (1978)
3. Brualdi, R.A.: Combinatorial Matrix Classes. Encyclopedia of Mathematics and its Applications, vol. 108. Cambridge University Press, Cambridge (2006)
4. Crama, Y., Hammer, P.L. (eds.): Boolean Models and Methods in Mathematics, Computer Science, and Engineering, 1st edn. Cambridge University Press (2010)
5. Emets', O.O., Emets', E.M.: Cut-off in linear partially combinatorial problems of Euclidean combinatorial optimization. Dopovidi Natsionalnoi Akademii Nauk Ukrainy. Matematika. Prirodoznavstvo. Tekhnichni Nauki (9), 105–109 (2000)
6. Gimadi, E., Khachay, M.: Extremal Problems on Sets of Permutations. UMC UPI, Ekaterinburg (2016). (in Russian)
7. Gmys, J.: Heterogeneous cluster computing for many-task exact optimization - application to permutation problems. phdthesis, Université de Mons (UMONS); Université de Lille, December 2017. https://hal.inria.fr/tel-01652000/document
8. Grande, F.: On k-level matroids: geometry and combinatorics. Doctor of Natural Sciences Dissertation, Institut für Mathematik und Informatik, Freie Universität Berlin (2015). http://www.diss.fu-berlin.de/diss/receive/FUDISS_thesis_000000100434
9. Green, R.M.: Homology representations arising from the half cube, II. J. Comb. Theory. Ser. A **117**(8), 1037–1048 (2010)
10. Grotschel, M., Lovasz, L., Schrijver, A.: Geometric Algorithms and Combinatorial Optimization. Algorithms and Combinatorics, 2nd edn. Springer, Heidelberg (1993). https://doi.org/10.1007/978-3-642-78240-4
11. Iemets, O.O., Yemets, Y.M., Parfionova, T.A., Chilikina, T.V.: Solving linear conditional completely combinatorial optimization problems on permutations by the branch and bound method. Cybern. Syst. Anal. **49**(2), 264–278 (2013). https://doi.org/10.1007/s10559-013-9508-1
12. Kellerer, H., Pferschy, U., Pisinger, D.: Knapsack Problems, 1st edn. Springer, Berlin (2004). https://doi.org/10.1007/978-3-540-24777-7. Softcover reprint of hardcover
13. Koliechkina, L.N., Dvernaya, O.A., Nagornaya, A.N.: Modified coordinate method to solve multicriteria optimization problems on combinatorial configurations. Cybern. Syst. Anal. **50**(4), 620–626 (2014). https://doi.org/10.1007/s10559-014-9650-4
14. Mehdi, M.: Parallel hybrid optimization methods for permutation based problems. Ph.D.thesis, Université des Sciences et Technologie de Lille - Lille I, Lille, October 2011. https://tel.archives-ouvertes.fr/tel-00841962/document
15. Pichugina, O., Kartashov, O.: Signed permutation polytope packing in VLSI design. In: 2019 IEEE 15th International Conference on the Experience of Designing and Application of CAD Systems (CADSM) Conference Proceedings, pp. 4/50–4/55. Lviv (2019). https://doi.org/10.1109/CADSM.2019.8779353
16. Pichugina, O., Yakovlev, S.: Continuous approaches to the unconstrained binary quadratic problems. In: Bélair, J., Frigaard, I.A., Kunze, H., Makarov, R., Melnik, R., Spiteri, R.J. (eds.) Mathematical and Computational Approaches in Advancing Modern Science and Engineering, pp. 689–700. Springer, Cham (2016). https://doi.org/10.1007/978-3-319-30379-6_62

17. Pichugina, O., Yakovlev, S.: Convex extensions and continuous functional representations in optimization, with their applications. J. Coupled Syst. Multiscale Dyn. **4**(2), 129–152 (2016). https://doi.org/10.1166/jcsmd.2016.1103

18. Pichugina, O., Yakovlev, S.: Euclidean combinatorial configurations: typology and applications. In: 2019 IEEE 2nd Ukraine Conference on Electrical and Computer Engineering (UKRCON 2019) Conference Proceedings, pp. 1065–1070. Lviv (2019)

19. Pichugina, O.S.: Surface and combinatorial cuttings in Euclidean combinatorial optimization problems. Math. Comput. Model. Ser.: Phys. Math. **1**(13), 144–160 (2016)

20. Pichugina, O., Muravyova, N.: A spherical cutting-plane method with applications in multimedia flow management. In: Proceedings of the 1st International Workshop on Digital Content & Smart Multimedia (DCSMart 2019), pp. 82–93. CEUR Vol-2533, Lviv, Ukraine, December 2019

21. Pichugina, O., Yakovlev, S.: Euclidean combinatorial configurations: continuous representations and convex extensions. In: Lytvynenko, V., Babichev, S., Wójcik, W., Vynokurova, O., Vyshemyrskaya, S., Radetskaya, S. (eds.) ISDMCI 2019. AISC, vol. 1020, pp. 65–80. Springer, Cham (2020). https://doi.org/10.1007/978-3-030-26474-1_5

22. Pogorelov, A.V.: Extrinsic Geometry of Convex Surfaces, 1st edn. American Mathematical Society, Providence (1973)

23. Shor, N.Z., Stetsyuk, P.I.: The use of a modification of the R-algorithm for finding the global minimum of polynomial functions. Cybern. Syst. Anal. **4**, 28–49 (1997). https://doi.org/10.1007/BF02733104

24. Stoyan, Y.G., Yakovlev, S.V.: Mathematical models and optimization methods in Geometric Design. Naukova Dumka, Kiev (1986)

25. Stoyan, Y.G., Yakovlev, S.V.: Theory and methods of Euclidian combinatorial optimization: current status and prospects. Cybern. Syst. Anal. **56**(3), 366–379 (2020). https://doi.org/10.1007/s10559-020-00253-6

26. Stoyan, Y.G., Yemets', O.: Theory and methods of Euclidean combinatorial optimization. ISSE, Kiev (1993). (in Ukrainian)

27. Stoyan, Y.G., Yakovlev, S.V., Parshin, O.V.: Quadratic optimization on combinatorial sets in RN. Cybern. Syst. Anal. **27**(4), 561–567 (1991). https://doi.org/10.1007/BF01130367

28. Yakovlev, S.: Convex extensions in combinatorial optimization and their applications. In: Butenko, S., Pardalos, P.M., Shylo, V. (eds.) Optimization Methods and Applications. SOIA, vol. 130, pp. 567–584. Springer, Cham (2017). https://doi.org/10.1007/978-3-319-68640-0_27

29. Yemelichev, V.A., Kovalev, M.M., Kravtsov, M.K.: Polytopes, Graphs and Optimisation. Cambridge University Press, Cambridge (1984)

Optimal Control

Operator Forms of the Maximum Principle and Iterative Algorithms in Optimal Control Problems

Alexander Buldaev[(✉)] [iD]

Buryat State University, Smolin Street, 24a, 670000 Ulan-Ude, Russia
buldaev@mail.ru

Abstract. The operator equations of the maximum principle are constructed in nonlinear optimal control problems in the form of fixed point problems in the control space. The equivalence of operator equations to the condition of the maximum principle is shown. The constructed operator forms of the maximum principle make it possible to apply and modify the well-known apparatus of the theory and methods of fixed points to search for extreme controls. The control operators under consideration define new iterative algorithms for finding extreme controls. The proposed iterative algorithms of fixed points of the maximum principle have the property of nonlocality of successive control approximations and the absence of a parametric procedure for improving the approximation at each iteration, which is characteristic of the well-known standard gradient type methods.

Keywords: Controllable system · Operator of control · Maximum principle · Fixed point problem · Iterative algorithm

1 Introduction

A well-known approach for solving optimal control problems is the construction and solution of systems of necessary conditions for optimal control. In particular, they construct and solve the boundary value problem of the maximum principle [1,2]. Another method is to build relaxation control sequences based on the sequential solution of problems of local control improvement. Under certain conditions, such sequences converge to extreme controls, i.e. satisfying the necessary conditions for optimality. An example of this approach is the well-known gradient methods [1–3].

The article considers a new approach to the search for extremal controls, which consists of constructing the necessary conditions for optimality of control in the form of operator equations, interpreted as fixed point problems in the control space. This form allows you to apply and adapt the theory and algorithms known in computational mathematics [4] to search for fixed points of

---

The work was supported by RFBR, project No. 18-41-030005.

N. Olenev et al. (Eds.): OPTIMA 2020, CCIS 1340, pp. 101–112, 2020.
https://doi.org/10.1007/978-3-030-65739-0_8

constructed operator equations. The fixed-point operator approach is illustrated in the framework of the class of optimal control problems with a free right end. The constructed operator forms of the maximum principle allow one to construct new iterative algorithms for searching for extreme controls. In [5–7], fixed point problems and methods were proposed based on the operation on the maximum of the Pontryagin function. In this paper, the fixed point approach under consideration is supplemented by new operator equations of the maximum principle based on maximum operators and projection operators.

2   Optimal Control Problem with a Free Right End

To illustrate the possibility of constructing new operator forms of necessary control optimality conditions, we consider the well-known classical optimal control problem with piecewise continuous controls [1–3], for which new forms of optimality conditions and the proposed fixed-point approach have a simple description.

The optimal control problem is considered:

$$\Phi(u) = \varphi(x(t_1)) + \int_T F(x(t), u(t), t)dt \rightarrow \inf_{u \in V}, \tag{1}$$

$$\dot{x}(t) = f(x(t), u(t), t), x(t_0) = x^0, u(t) \in U, t \in T = [t_0, t_1], \tag{2}$$

where $x(t) = (x_1(t), ..., x_n(t))$ – state of a system, $u(t) = (u_1(t), ..., u_m(t))$ – control. The set of admissible controls consists of piecewise continuous functions taking values in a convex compact set $U \subset R^m$:

$$V = \{v \in PC(T) : v(t) \in U, t \in T\}.$$

The initial state x^0 and time interval T are set.

The following notation is used: q_x – partial derivative of the first order of the function q for the corresponding argument x; $\langle x, y \rangle = \sum_{i=1}^{n} x_i y_i$ – a scalar product of vectors x,y in Euclidean space E^n; $\|x\|$ – norm of a vector in Euclidean space.

It is assumed that the function $\varphi(x)$ is continuously differentiable on R^n; functions $F(x, u, t)$, $f(x, u, t)$ and their derivatives $F_x(x, u, t)$, $F_u(x, u, t)$, $f_x(x, u, t)$, $f_u(x, u, t)$ are continuous in the totality of arguments on the set $R^n \times U \times T$; the function $f(x, u, t)$ satisfies the Lipschitz condition for x in $R^n \times U \times T$ with a constant $L > 0$:

$$\|f(x, u, t) - f(y, u, t)\| \leq L \|x - y\|.$$

Consider the Pontryagin function with an adjoint variable $\psi \in R^n$

$$H(\psi, x, u, t) = \langle f(x, u, t), \psi \rangle - F(x, u, t).$$

The standard conjugate system has the form:

$$\dot{\psi}(t) = -H_x(\psi(t), x(t), u(t), t), \psi(t_1) = -\varphi_x(x(t_1)).$$

For an admissible control $v \in V$, we denote $x(t, v)$, $t \in T$ – the solution of system (2); $\psi(t, v), t \in T$ – adjoint system solution for $u(t) = v(t)$, $x(t) = x(t, v)$, $t \in T$.

Using function

$$u^*(\psi, x, t) = \arg\max_{w \in U} H(\psi, x, w, t), \psi \in R^n, x \in R^n, t \in T. \tag{3}$$

the well-known necessary condition for optimal control (maximum principle) [1–3] can be represented in the following form:

$$v(t) = u^*(\psi(t, v), x(t, v), t), t \in T. \tag{4}$$

Relation (4) on the set of admissible controls is equivalent to the boundary value problem of the maximum principle in the state space:

$$\dot{x}(t) = f(x(t), u^*(\psi(t), x(t), t), t), x(t_0) = x^0, \tag{5}$$

$$\dot{\psi}(t) = -H_x(\psi(t), x(t), u^*(\psi(t), x(t), t), t), \psi(t_1) = -\varphi_x(x(t_1)). \tag{6}$$

Equivalence is understood in the following sense.

Let a pair $(x(t), \psi(t))$, $t \in T$ be a solution to the boundary value problem (5), (6). Then the output control generated by rule (3) $v(t) = u^*(\psi(t), x(t), t)$ satisfies condition (4). Conversely, let control $v \in V$ be a solution to a problem (4). Then the formed pair of functions $(x(t, v), \psi(t, v)), t \in T$ by their definition, satisfies the boundary value problem (5), (6).

In the general case, the right-hand sides of the boundary-value problem (5), (6) are discontinuous and multi-valued in phase variables x, ψ.

The maximum principle (4) implies the weakened necessary optimality condition, known as the differential maximum principle [2,3], which can be represented in the form of inequality:

$$\langle H_u(\psi(t, u), x(t, u), u(t), t), w - u(t) \rangle \le 0, w \in U, t \in T. \tag{7}$$

We introduce the map w^α, $\alpha > 0$ using the relation

$$w^\alpha(\psi, x, u, t) = P_U(u + \alpha H_u(\psi, x, u, t)), \psi \in R^n, x \in R^n, u \in U, t \in T,$$

where P_U is the set U projection operator in the Euclidean norm

$$P_U(z) = \arg\min_{w \in U}(\|w - z\|), z \in R^m.$$

Based on the Lipschitz condition for the operator P_U the function w^α is continuous in the aggregate. $(\psi, x, u, t) \in R^n \times R^n \times U \times T$.

The differential maximum principle (7) for control $u \in V$ through a mapping w^α can be written in the following form:

$$u(t) = w^\alpha(\psi(t, u), x(t, u), u(t), t), t \in T, \alpha > 0. \tag{8}$$

Note that for (7) to be satisfied, it suffices to check condition (8) for at least one $\alpha > 0$. Conversely, condition (7) implies the fulfillment of (8) for all $\alpha > 0$.

We single out an important for applications subclass of linear control problems (functions $f(x, u, t)$, $F(x, u, t)$ are linear in u). In a linear control problem, the differential maximum principle (7) is equivalent to the maximum principle (4), and to search for controls satisfying the maximum principle, one can use the projection form (8) of the differential maximum principle, which is simpler in terms of smoothness than the condition of the maximum principle (4).

3   Operator Equations Based on a Maximum Operation

The condition of the maximum principle (4) can be interpreted as the problem of a fixed point of some control operator:

$$v = G_1^*(v), v \in V.$$

in which the operator G_1^* can be formalized as a superposition of three mappings.

The first mapping X is defined using the relation

$$X(v) = x, v \in V, x(t) = x(t, v), t \in T.$$

We construct the second map Ψ in the same way:

$$\Psi(v) = \psi, v \in V, \psi(t) = \psi(t, v), t \in T.$$

We construct the third mapping V^* in the form

$$V^*(\psi, x) = v^*, \psi \in C(T), x \in C(T), v^*(t) = u^*(\psi(t), x(t), t), t \in T,$$

where $C(T)$ is the space of continuous functions on T.

As a result, problem (4) can be represented as an operator equation in the control space:

$$v = V^*(\Psi(v), X(v)), v \in V. \tag{9}$$

Equation (9) can be written in the canonical form of the problem of a fixed point with an operator G_1^* defined as a superposition:

$$G_1^*(v) = V^*(\Psi(v), X(v)).$$

We construct new operator problems on a fixed point equivalent to the boundary value problem of the maximum principle (5), (6), and condition (4).

We introduce the mapping X^* as follows:

$$X^*(\psi) = x, \psi \in C(T), x \in C(T),$$

where $x(t), t \in T$ is the solution to the special Cauchy problem

$$\dot{x}(t) = f(x(t), u^*(\psi(t), x(t), t), t), x(t_0) = x^0.$$

Consider the operator equation

$$v = V^*(\Psi(v), X^*(\Psi(v))), v \in V. \tag{10}$$

Indeed, let $v \in V$ be a solution to Eq. (9), i.e. a pair $(x(t, v), \psi(t, v))$, $t \in T$ is a solution to the boundary value problem (5), (6).

Then the function $x(t, v), t \in T$ is a solution to the Cauchy problem

$$\dot{x}(t) = f(x(t), u^*(\psi(t, v), x(t), t), t), x(t_0) = x^0,$$

i.e. $X(v) = X^*(\Psi(v))$. Hence we get that

$$V^*(\Psi(v), X^*(\Psi(v))) = V^*(\Psi(v), X(v)) = v.$$

Conversely, let $v \in V$ be a solution to Eq. (10), i.e.

$$v(t) = u^*(\psi(t, v), x(t), t),$$

where $x(t), t \in T$ is the solution to the special Cauchy problem

$$\dot{x}(t) = f(x(t), u^*(\psi(t, v), x(t), t), t), x(t_0) = x^0.$$

Hence, $x(t) = x(t, v), t \in T$, i.e. $X^*(\Psi(v)) = X(v)$. Thus we get:

$$V^*(\Psi(v), X(v)) = V^*(\Psi(v), X^*(\Psi(v))) = v.$$

Consider a control operator G_2^* in the form of a superposition of mappings:

$$G_2^*(v) = V^*(\Psi(v), X^*(\Psi(v))).$$

Then the operator Eq. (10) is represented in the form of the canonical problem of a fixed point:

$$v = G_2^*(v), v \in V.$$

In pointwise form, problem (10) can be written as:

$$v(t) = u^*(\psi(t, v), x(t, V^*(\Psi(v), X^*(\Psi(v)))), t), t \in T.$$

We obtain another operator problem of a fixed point equivalent to the boundary value problem of the maximum principle and condition (4) using the following mapping:

$$\Psi^*(x) = \psi, x \in C(T), \psi \in C(T),$$

in which $\psi(t), t \in T$ is a solution to the special conjugate Cauchy problem

$$\dot{\psi}(t) = -H_x(\psi(t), x(t), u^*(\psi(t), x(t), t), t), \psi(t_1) = -\varphi_x(x(t_1)).$$

Consider the operator equation

$$v = V^*(\Psi^*(X(v)), X(v)), v \in V. \tag{11}$$

Similarly to the above argument, one can show the equivalence of Eqs. (11) and (9).

We construct the control operator G_3^* by the formula:

$$G_3^*(v) = V^*(\Psi^*(X(v)), X(v)).$$

Then Eq. (11) is represented in the canonical form of the fixed point problem

$$v = G_3^*(v), v \in V.$$

In pointwise form, problem (11) is written as:

$$v(t) = u^*(\psi(t, V^*(\Psi^*(X(v)), X(v))), x(t, v), t), t \in T.$$

Thus, based on the above reasoning, the following statement can be formulated.

Theorem 1. *The operator Eqs. (9), (10), (11) are equivalent to the condition of the maximum principle (4).*

4   Operator Equations Based on Projection Operation

The condition of the differential maximum principle in projection form (8) can be represented in the form of equivalent operator equations on the set of admissible controls, interpreted as fixed point problems.

We introduce the auxiliary operator $V^\alpha, \alpha > 0$ by the relation

$$V^\alpha(\psi, x, v) = v^\alpha, \psi \in C(T), x \in C(T), v \in V,$$

$$v^\alpha(t) = w^\alpha(\psi(t), x(t), v(t), t) = P_U(v(t) + \alpha H_u(\psi(t), x(t), v(t), t)), t \in T.$$

Define the operator $X^\alpha, \alpha > 0$:

$$X^\alpha(\psi, v) = x^\alpha, \psi \in C(T), v \in V, x^\alpha(t) = x^\alpha(t, \psi, v), t \in T,$$

where $x^\alpha(t, \psi, v), t \in T$ is the solution to the Cauchy problem:

$$\dot{x}(t) = f(x(t), w^\alpha(\psi(t), x(t), v(t), t), t), x(t_0) = x^0.$$

Build the operator $\Psi^\alpha, \alpha > 0$:

$$\Psi^\alpha(x, v) = \psi^\alpha, x \in C(T), v \in V, \psi^\alpha(t) = \psi^\alpha(t, x, v),$$

where $\psi^\alpha(t, x, v), t \in T$ is the solution of the conjugate Cauchy problem:

$$\dot{\psi}(t) = -H_x(\psi(t), x(t), w^\alpha(\psi(t), x(t), v(t), t), t), \psi(t_1) = -\varphi_x(x(t_1)).$$

Based on the mappings introduced earlier $\Psi : u \to \psi(t, u), t \in T$ and $X : u \to x(t, u), t \in T$ we construct the operators $G_1^\alpha, G_2^\alpha, G_3^\alpha$ in the form:

$$G_1^\alpha(v) = V^\alpha(\Psi(v), X(v), v), v \in V,$$

$$G_2^\alpha(v) = V^\alpha(\Psi(v), X^\alpha(\Psi(v), v), v), v \in V,$$
$$G_3^\alpha(v) = V^\alpha(\Psi^\alpha(X(v), v), X(v), v), v \in V.$$

We consider three operator equations in the form of fixed point problems

$$v = V^\alpha(\Psi(v), X(v), v) = G_1^\alpha(v), v \in V, \alpha > 0, \tag{12}$$

$$v = V^\alpha(\Psi(v), X^\alpha(\Psi(v), v), v) = G_2^\alpha(v), v \in V, \alpha > 0, \tag{13}$$

$$v = V^\alpha(\Psi^\alpha(X(v), v), X(v), v) = G_3^\alpha(v), v \in V, \alpha > 0. \tag{14}$$

Similarly to the previous section, the following statement can be obtained.

Theorem 2. *The operator Eqs. (12), (13), (14) are equivalent to the condition of the differential maximum principle (8).*

5   Iterative Algorithms Based on Maximum Operators

The search for extremal controls satisfying the necessary conditions for optimality of control (9)–(11) and (12)–(13) can be considered as a search for solutions to the corresponding fixed point problems in the control space. A well-known approach to analyzing the existence of solutions to fixed point problems in a Banach space is the construction of iterative processes converging to solving fixed point problems. In this case, the conditions for the convergence of iterative processes are determined based on the well-known principle of squeezed mappings.

As an example, consider the fixed point problem:

$$v = G(v), v \in V_E, \tag{15}$$

in which $G : V_E \to V_E$ is an operator acting on a set V_E in a full normed space E with norm $\| \cdot \|_E$.

For the numerical solution of problem (15) one can consider the method of simple iteration at $k \geq 0$, having the form:

$$v^{k+1} = G(v^k), v^0 \in V_E \tag{16}$$

The conditions for the convergence of the iterative process (16) to the solution of problem (15) can be easily obtained similarly to [4].

A similar analysis of the existence of solutions to fixed point problems (9)–(11) and (12)–(13) based on this approach should also be carried out in Banach control spaces.

Analysis of solutions to the problem (9)–(11) can be considered in a wider Banach space of measurable functions:

$$V \subset V_L = \{v \in L_\infty(T) : v(t) \in U, t \in T\}$$

with the norm $\|v\|_\infty = \operatorname{ess\,sup}_{t \in T} \|v(t)\|, v \in V_L$. For this, it is necessary to generalize the optimal control problem (1), (2) by extending the set of piecewise continuous controls to the specified set of measurable controls.

The corresponding methods of simple iteration for finding solutions to problems (9)–(11) have the following form for $k \geq 0$:

$$v^{k+1} = V^*(\Psi(v^k), X(v^k)), v^0 \in V_L, \tag{17}$$

$$v^{k+1} = V^*(\Psi(v^k), X^*(\Psi(v^k))), v^0 \in V_L, \tag{18}$$

$$v^{k+1} = V^*(\Psi^*(X(v^k)), X(v^k)), v^0 \in V_L. \tag{19}$$

In pointwise form, the first method has the form:

$$v^{k+1}(t) = u^*(\psi(t, v^k), x(t, v^k), t), v^0 \in V_L, t \in T.$$

According to the definition of mappings, the following relation holds:

$$X(V^*(\Psi(v), X^*(\Psi(v)))) = X^*(\Psi(v)), v \in V_L. \tag{20}$$

Indeed, for anyone $p \in C(T)$ by definition we get:

$$X^*(p)|_t = x(t), t \in T,$$

where $x(t), t \in T$ is a solution to the Cauchy problem:

$$\dot{x}(t) = f(x(t), u^*(p(t), x(t), t), t), x(t_0) = x^0.$$

Further, according to the definition, we have:

$$V^*(p, X^*(p))|_t = u^*(p(t), x(t), t), t \in T,$$

where $x(t), t \in T$ is a solution to the Cauchy problem:

$$\dot{x}(t) = f(x(t), u^*(p(t), x(t), t), t), x(t_0) = x^0.$$

Hence,

$$x(t) = X^*(V^*(p, X^*(p)))|_t, t \in T.$$

Thus, from pointwise equalities we obtain the operator equality:

$$X(V^*(p, X^*(p))) = X^*(p), p \in C(T),$$

from which follows (20).

According to the iterative process, from (20) it follows:

$$X^*(\Psi(v^k)) = X(V^*(\Psi(v^k), X^*(\Psi(v^k)))) = X(v^{k+1}).$$

Therefore, the second method of simple iteration (18) is presented in the following implicit form:

$$v^{k+1} = V^*(\Psi(v^k), X(v^{k+1})), v^0 \in V_L, \tag{21}$$

or in pointwise form:

$$v^{k+1}(t) = u^*(\psi(t, v^k), x(t, v^{k+1}), t), v^0 \in V_L, t \in T$$

To evaluate the computational efficiency of iterative algorithms, it is important to note that the complexity of implementing one iteration of implicit methods (18), (19) is similar to the complexity of implementing the explicit method (17) and consists of two Cauchy problems for phase and conjugate variables.

Indeed, at the k-th iteration in the process (18) after calculating the solution of the Cauchy problem $\psi(t, v^k), t \in T$ the solution of the phase system $x(t), t \in T$ is found:

$$\dot{x}(t) = f(x(t), u^*(\psi(t, v^k), x(t), t), t), x(t_0) = x^0.$$

Then the output control is built according to the rule:

$$v^{k+1}(t) = u^*(\psi(t, v^k), x(t), t), t \in T.$$

Moreover, by virtue of the construction, the relation is satisfied:

$$x(t) = x(t, v^{k+1}), t \in T.$$

Similarly, at the k-th iteration of the process (19) after calculation $x(t, v^k), t \in T$ a solution $\psi(t), t \in T$ is found for the conjugate system:

$$\dot{\psi}(t) = -H_x(\psi(t), x(t, v^k), u^*(\psi(t), x(t, v^k), t), t), \psi(t_1) = -\varphi_x(x(t_1, v^k)).$$

Then the output control is built according to the rule:

$$v^{k+1}(t) = u^*(\psi(t), x(t, v^k), t), t \in T,$$

for which, by construction, the relation holds:

$$\psi(t) = \psi(t, v^{k+1}), t \in T.$$

Note that only at the initial iteration of process (18) with $k = 0$ for calculation $\psi(t, v^0), t \in T$ it is necessary to solve the additional Cauchy problem in order to obtain a solution $x(t, v^0), t \in T$.

Comparing the proposed algorithms with other well-known iterative methods of the maximum principle, we note that the method (17) is equivalent to the simplest method of successive approximations [8]. No known analogs of iterative fixed-point methods (18) and (19) have been found in the literature.

To comparatively highlight the characteristic features of the proposed fixed-point operator methods (17)–(19), we consider the structure of two common known maximum principle methods in the notation used.

The standard method of conditional gradient [2,3] is described by the relations:

$$\bar{v}^k(t) = u^*(\psi(t, v^k), x(t, v^k), t), t \in T, v^0 \in V, k \geq 0,$$

$$v_k^\lambda(t) = v^k(t) + \lambda(\bar{v}^k(t) - v^k(t)), t \in T,$$

$$\lambda \in [0, 1] : \Phi(v_\lambda^k) \leq \Phi(v^k) \Rightarrow v^{k+1}(t) = v_\lambda^k(t), t \in T.$$

The needle linearization method [3] is characterized by the relations:

$$\bar{v}^k(t) = u^*(\psi(t, v^k), x(t, v^k), t), t \in T, v^0 \in V, k \geq 0,$$

$$g^k(t) = \Delta_{\bar{v}^k} H(\psi_1(t, v^k), x_1(t, v^k), v^k(t), t), t \in T,$$

$$\lambda_{min} = \inf_{t \in T} g^k(t), \lambda_{max} = \sup_{t \in T} g^k(t),$$

$$v_\lambda^k(t) = \begin{cases} v^k(t), g^k(t) \le \lambda, \\ \bar{v}^k(t), g^k(t) > \lambda, \end{cases}, \lambda \in [\lambda_{min}, \lambda_{max}], t \in T,$$

$$\lambda \in [\lambda_{min}, \lambda_{max}] : \Phi(v_\lambda^k) \le \Phi(v^k) \to v^{k+1}(t) = v_\lambda^k(t), t \in T.$$

A characteristic feature of these known methods is the search for a first approximation of control, which then varies in the vicinity of the improved control in order to improve the target functional of the problem.

Thus, in the proposed fixed-point operator methods, in contrast to the known gradient methods and maximum principle methods, relaxation by the objective functional at each iteration of the methods is not guaranteed. The nonlocality of successive control approximations and the absence of a rather laborious operation of convex or needle-shaped variation of the control in the vicinity of the current control are compensated for by the relaxation property.

6   Iterative Algorithms Based on Projection Operators

The search for solutions to problems (12)–(14) can be investigated in a narrower Banach space of continuous controls:

$$V_C = \{v \in C(T) : v(t) \in U, t \in T\} \subset V$$

with the norm $\|v\|_C = \max_{t \in T} \|v(t)\|, v \in V_C$.

Such a narrowing of the solution search space is admissible if converging iterative processes are constructed in the class of continuous admissible controls.

Simple iteration methods for solving problems (12)–(14) in the space of continuous controls have the following form for $k \ge 0$:

$$v^{k+1} = V^\alpha(\Psi(v^k), X(v^k), v^k), v^0 \in V_C, \alpha > 0, \tag{22}$$

$$v^{k+1} = V^\alpha(\Psi(v^k), X^\alpha(\Psi(v^k), v^k), v^k), v^0 \in V_C, \alpha > 0, \tag{23}$$

$$v^{k+1} = V^\alpha(\Psi^\alpha(X(v^k), v^k), X(v^k), v^k), v^0 \in V_C, \alpha > 0. \tag{24}$$

It can be easily shown, due to the properties of the design operation, that if $v^0 \in V_C$, then successive approximations of the control for $k > 0$ will also be continuous controls.

In a point form, the iterative method (22) takes the form:

$$v^{k+1}(t) = w^\alpha(\psi(t, v^k), x(t, v^k), v^k(t), t), v^0 \in V_C, \alpha > 0, t \in T.$$

Similarly to obtaining relation (20) in the previous section, we can obtain the following operator relation:

$$X(V^\alpha(p, X^\alpha(p, v), v)) = X^\alpha(p, v), p \in C(T), v \in V_C.$$

From here we have:

$$X^\alpha(\Psi(v^k), v^k) = X(V^\alpha(\Psi(v^k), X^\alpha(\Psi(v^k), v^k), v^k)) = X(v^{k+1}).$$

Thus, the second method of simple iteration (23) to search for fixed points of the differential maximum principle can be written in the implicit form:

$$v^{k+1} = V^\alpha(\Psi(v^k), X(v^{k+1}), v^k), v^0 \in V_C, \alpha > 0.$$

In a point form, iterative methods of the differential maximum principle takes the form:

$$v^{k+1}(t) = w^\alpha(\psi(t, v^k), x(t, v^k), v^k(t), t), v^0 \in V_C, \alpha > 0, t \in T.$$

The complexity of the computational implementation of one iteration of explicit and implicit projection methods (22)–(24) consists of two Cauchy problems for phase and conjugate variables.

Similarly to [5], we can formulate simple conditions for the convergence of iterative processes (22)–(24) with continuous initial approximations to continuous solutions of the corresponding fixed point problems for sufficiently small projection parameters $\alpha > 0$.

No well-known analogs of projection iterative fixed-point methods (22)–(24) were found in the literature.

To compare the developed projection methods of fixed points, we will present in the notation used the standard gradient projection method with $\alpha > 0$ [2,3]:

$$v_\alpha^k(t) = w^\alpha(\psi(t, v^k), x(t, v^k), v^k(t), t), t \in T, v^0 \in V, k \geq 0,$$

$$\alpha \in (0, \infty) : \Phi(v_\alpha^k) \leq \Phi(v^k) \Rightarrow v^{k+1} = v_\alpha^k.$$

At each iteration of the gradient gradient projection method under consideration, the projection parameter is varied to provide improved control.

In the constructed fixed-point projection methods, in contrast to the standard gradient projection method, the design parameter $\alpha > 0$ is fixed in the iterative process of successive control approximations. Thus, at each iteration of the proposed methods, relaxation with respect to the objective functional is not guaranteed, but this property is compensated by the nonlocality of successive control approximations, the absence of the operation of varying control in the vicinity of the current approximation to provide an improvement in the functional.

7 Conclusion

The main result of this work is to obtain new operator forms of known necessary optimality conditions in the considered class of optimal control problems. The obtained operator forms can be interpreted as fixed point problems and allow developing a new approach to the search for extremal controls, which consists of constructing iterative algorithms for solving the indicated fixed point problems.

The proposed new fixed point approach for finding extreme controls is characterized by the following main features.

1. Nonlocality of successive control approximations in constructed iterative processes for searching for extremal controls.
2. Absence of a laborious procedure of needle or convex variation of control at each iteration of successive approximations, which is typical for gradient control methods.
3. The computational complexity of each iteration of successive control approximations is estimated by solving two Cauchy problems for phase and conjugate variables.
4. Computational stability of the calculation of fixed point problems, which is determined by a separate calculation of alternating phase and conjugate systems.

The indicated properties of the proposed approach for the search for extreme controls are important for increasing the efficiency of solving optimal control problems and determine the direction of developing new methods for optimizing controlled systems.

References

1. Vasilyev, F.P.: Chislennye metody resheniya ekstremalnykh zadach [Numerical Methods for Solving Extremal Problems]. Nauka Publ., Moscow (1980)
2. Vasilyev, O.V.: Lektsii po metodam optimizatsii [Lectures on Optimization Methods]. Irkutsk State University Publ., Irkutsk (1994)
3. Srochko, V.A.: Iteratsionnye metody resheniya zadach optimalnogo upavleniya [Iterative Methods for Solving Optimal Control Problems]. Fizmatlit Publ., Moscow (2000)
4. Samarskii, A.A., Gulin, A.V.: Chislennye metody [Numerical Methods]. Nauka Publ., Moscow (1989)
5. Buldaev, A.S.: Metody vozmushchenii v zadachakh uluchsheniya i optimizatsii upravlyaemykh system [Perturbation Methods in the Problems of Improving and Optimizing Controllable Systems]. Buryat State University Publ., Ulan-Ude (2008)
6. Buldaev, A.S.: Metody nepodvizhnykh tochek printsipa maksimuma [Methods of Fixed Points of the Maximum Principle]. Vestnik Buryatskogo gosudarstvennogo universiteta. Matematika, informatika (4), 36–46 (2015)
7. Buldaev, A.S.: Zadachi i metody nepodvizhnykh tochek printsipa maksimuma [Problems and Methods of Fixed Points of the Maximum Principle]. Izvestiya Irkutskogo gosudarstvennogo universiteta. Seriya Matematika **14**, 31–41 (2015)
8. Chernousko, F.L.: Otsenivanie fazovogo sostoyaniya dinamicheskikh system [Derivation of Estimate for the Phase State of Dynamical Systems]. Nauka Publ., Moscow (1988)

Solution of the Problem of the Control System General Synthesis by Approximation of a Set of Extremals

Askhat Diveev[1,2](✉) [ID] and Sergey Konstantinov[1,2] [ID]

[1] Federal Research Center "Computer Science and Control"
of Russian Academy of Sciences, Moscow 119333, Russia
aidiveev@mail.ru
[2] Peoples' Friendship University of Russia (RUDN University),
Moscow 117198, Russia

Abstract. The problem of the control system general synthesis is considered. This problem in general case requires finding the solution in the form of multidimensional function of a vector argument. Placing this control function into the right-hand part of differential equations of the control object model allows receiving the system of differential equations which partial solution from any initial condition of the given set is always optimal trajectory for the given quality criterion. In this paper, the problem of control general synthesis is solved based on the approximation of the set of optimal control problem solutions for different initial conditions. These solutions are called extremals. Previously, to solve the general synthesis problem, symbolic regression methods were used without approximation of extremals. Therefore it was often impossible to estimate the proximity of the found solution to the optimal one. To avoid this issue in this work initially we solve the optimal control problems for different initial conditions, and then these solutions are approximated by the symbolic regression method. In a presented computational experiment the proposed approach is used to solve the problem of the control system general synthesis for the mobile robot moving in the area with obstacles.

Keywords: Control synthesis · Optimal control · Extremals ·
Evolutionary algorithms · Symbolic regressions.

1 Introduction

The problem of control general synthesis is a very complex one for the numerical solution. The main difficulty lies in a form of solution. The solution is a multidimensional function of the state space vector as an argument [1]. By replacing

---

The theoretical part of the research, Sections. 1–4, was supported by the Russian Science Foundation (project No 19-11-00258). Experimental and computational part of the research, Section 5, was supported by the Russian Foundation for Basic Research (project No 18-29-03061-mk).

© Springer Nature Switzerland AG 2020
N. Olenev et al. (Eds.): OPTIMA 2020, CCIS 1340, pp. 113–128, 2020.
https://doi.org/10.1007/978-3-030-65739-0_9

control vector with this function in the right-hand part of differential equations of the control object mathematical model the system of differential equations is obtained each partial solution of which for any initial conditions from the given domain is a solution of the optimal control problem. Such function may be non differentiable and may have a discontinuity of the first kind. The searching of the problem solution in the class of differentiable continuous functions contradicts the practical control systems. Often the controls for a real object can be discontinuous and not smooth.

Previously, in regression problems, the researcher defined the required function accurate to parameter values. Then, optimization algorithms were used to search the optimal values of these parameters. Last decade for solving the problem of the control system general synthesis symbolic regression methods are used [2,3]. Symbolic regression methods allow to search for the structure of a mathematical expression. These methods encode the possible solution, which in our case is the mathematical expression for the searched control function, and search for the optimal code by some evolutionary algorithms [4], in most cases by a genetic algorithm which search for a mathematical expression on a set of codes [5]. All symbolic regression methods are differed by the form of codes. The main operations of genetic algorithm like crossover and mutation depend on coding rules of the specific symbolic regression method. At the moment there are over 10 different symbolic regression methods.

The main drawback of direct approaches for solving the problem of optimal control general synthesis based on symbolic regression methods is that genetic algorithm doesn't provide an information about the proximity of a found solution to the optimal one. The solution found by symbolic regression method is the best one from all checked possible solutions during the search process. Note that in order to check one possible solution it is necessary to place it in the right-hand part of system of differential equations of control object model and integrate this system for all given initial conditions.

To eliminate this defect we suggest a two-step approach to solve the problem of the control system general synthesis. At the first step we solve the optimal control problem for all given initial conditions. As a result we receive a set of extremals that are discretely stored in a computer memory as control vectors and state space vectors in some time moments. As the result we got a set of points in the state space for each optimal control problem. At the second step we approximate these solutions by symbolic regression method [6]. So the value of quality criterion for the approximation is an estimation of proximity of the found solution to the optimal one.

Unlike previous approaches for solving the problem of optimal control general synthesis using symbolic regression methods, the suggested approach allows to estimate the proximity of the found solution to the optimal one in terms of the value of quality criterion, since for each initial condition we have a previously solved optimal control problem with the corresponding quality criterion value.

In this paper the problem of the control system general synthesis statement and the numerical method for its solution based on the approximation of the extremals are described. The optimal control problem statement and some evolutionary algorithms for its direct solution are presented. Next a symbolic regression method is described briefly and computational experiment of solving the problem of the control system general synthesis using proposed approach is presented.

2   Statement of the Problem of the Control System General Synthesis

A mathematical model of the control object is given

$$\dot{\mathbf{x}} = \mathbf{f}(\mathbf{x}, \mathbf{u}), \tag{1}$$

where $\mathbf{x}$ is a state space vector, $\mathbf{x} \in \mathbb{R}^n$, $\mathbf{u}$ is a control vector, $\mathbf{u} \in U \subseteq \mathbb{R}^m$, U is bounded closed set, $m \leq n$.

A domain of initial conditions is

$$X_0 \subseteq \mathbb{R}^n, \tag{2}$$

Terminal conditions are

$$\mathbf{x}(t_f(\mathbf{x}^0)) = \mathbf{x}^f \in \mathbb{R}^n, \tag{3}$$

where $t_f(\mathbf{x}^0)$ is a time of archiving the terminal conditions $\mathbf{x}^f$ from initial condition $\mathbf{x}^0 \in X_0$. $\forall \mathbf{x}^0 \in X_0$, $t_f(\mathbf{x}^0)$ is limited, $t_f(\mathbf{x}^0) \leq t^+$, t^+ is a given maximum time for archiving the terminal state (3). If a possible solution doesn't provide reaching the terminal state (3) in time t^+ then the control process forcibly ends.

A quality criterion is

$$J = \overbrace{\int_{X_0} \ldots \int}^{n} \int_0^{t_f(\mathbf{x}^0)} f_0(\mathbf{x}(t), \mathbf{u}(t)) dt dx_1 \ldots dx_n \to \min_{\mathbf{u} \in U}, \tag{4}$$

It is required to find a control as a function of state space vector

$$\mathbf{u} = \mathbf{h}(\mathbf{x}) \in U \tag{5}$$

such that any partial solution of the system of differential equations

$$\dot{\mathbf{x}} = \mathbf{f}(\mathbf{x}, \mathbf{h}(\mathbf{x})) \tag{6}$$

from the domain (2) achieves the terminal conditions (3) and provides an optimal value of the criterion (4).

In order to create a numerical computational algorithm for solving this problem we have to reformulate the problem statement. Let us replace the domain of initial conditions (2) with a finite set of initial conditions, then replace multiple integrals with a sum of integrals for each initial condition from a finite set and define the procedure of calculation of the terminal time of the control process

$$\tilde{X}_0 = \{\mathbf{x}^{0,1}, \ldots, \mathbf{x}^{0,M}\}, \tag{7}$$

$$\tilde{J} = \sum_{j=1}^{M} \int_0^{t_f(\mathbf{x}^{0,j})} f_0(\mathbf{x}(t), \mathbf{u}(t)) dt \to \min_{\mathbf{u} \in U}, \tag{8}$$

where

$$t_f(\mathbf{x}^{0,j}) = \begin{cases} t, \text{if } t < t^+ \text{and } \left\| \mathbf{x}^f - \mathbf{x}(t, \mathbf{x}^{0,j}) \right\| \le \varepsilon_0 \\ t^+, \text{otherwise} \end{cases}, \tag{9}$$

ε_0 is a small positive value, $\mathbf{x}(t, \mathbf{x}^{0,j})$ is a solution of the system of differential Eqs. (6) with initial condition $\mathbf{x}^{0,j} \in \tilde{X}_0$, $1 \le j \le K$.

The problem solution allows to receive an optimal control $\tilde{\mathbf{u}} = \mathbf{h}(\mathbf{x})$ and trajectory $\tilde{\mathbf{x}}(t, \mathbf{x}^0)$ for any initial condition from the given set $\forall \mathbf{x}^0 \in \tilde{X}_0$. If we are considering the control function as a function of time, $\tilde{\mathbf{u}} = \mathbf{h}(\mathbf{x}(t, \mathbf{x}^0))$, then this control function is the solution of the partial optimal control problem. Let us consider that the synthesis problem (1)–(8) is solved if the found control function (5) allows to determinate optimal solutions for all given initial conditions from the set (7) as well as for other initial conditions from the domain (2).

In our review we don't consider analytical and semi-analytical methods such as solution of the Bellman equation [7]. It is obvious that this won't suite for the most cases. Semi-analytical methods, e.g. the method of backstepping [8] or the method of analytical design of aggregated controllers [9], require some certain properties of a mathematical model and can be used only for providing the stability of a control object in the terminal state (3) without calculations of functionals (4) or (8).

The problem (1)–(8) can be solved only by numerical methods of symbolic regression. We assume that it also can be solved by an artificial neural network (ANN), but we don't know any related works with such solution. Moreover, the solution of this problem by ANN won't provide us with a mathematical equation for further analysis and studies.

Symbolic regression methods apply complex, compute-intensive procedures and allow to find solutions in the form of mathematical expression. Sometimes the mathematical expression found by these methods can be a more than a few dozen lines. But this is not the main weakness of symbolic regression methods. Solutions in these methods are found by evolutionary algorithms. The main drawback of symbolic regression methods and of all evolutionary algorithms in general is that we don't get the proximity of the found solution to the optimal one. Despite evolutionary algorithms can be considered as methods of random search, studies show that any evolutionary algorithm performs better than any random search algorithm [10]. If evolutionary algorithms are used in solving the optimization problem we believe that it is likely that term "training" should be used instead of the term "search".

In this work initially the optimal control problem is solved numerically for each initial condition from the set (7) forming a training data set. Then symbolic regression method is used to approximate the data set. Drawing analogies to neural networks technologies, this approach is similar to training with a teacher (supervised learning) in contrast to the direct solving the synthesis problem with the same symbolic regression method. A direct solution is similar to training without a teacher (unsupervised learning). A preliminary solution of multiple optimal control problems from different initial conditions for the synthesis problem can be named as a procedure of obtaining training set.

3   Solution of the Optimal Control Problem

Consider the optimal control problem statement. A mathematical model of a control object is given in (1).

There is only one initial condition

$$\mathbf{x}(0) = \mathbf{x}^0 \in \tilde{X}_0, \tag{10}$$

The terminal condition is given in (3).

The quality criterion is

$$J_1 = \int_0^{t_f} f_0(\mathbf{x}(t), \mathbf{u}(t))dt \to \min_{\mathbf{u} \in U}, \tag{11}$$

where terminal time t_f is determined by formula (9).

It is known that for a numerical solution of the optimal control problem there are two approaches: direct and indirect. The first approach is to transform the original optimization problem in infinite-dimensional space into a nonlinear programming problem. An indirect approach involves the application of the Pontryagin maximum principle, that modifies the optimal control problem into a boundary value problem. It has been shown in many computational experiments that applying the Pontryagin maximum principle makes sense for analytical solutions if they can be found. For numerical solutions the maximum principle doesn't give any advantages [11]. Therefore, we use a direct approach to solve the optimal control problem.

Set an interval on the time axis Δt. For a limited time t^+ we have no more than

$$K = \left\lceil \frac{t^+}{\Delta t} \right\rceil \tag{12}$$

intervals.

The required control function is represented in the form of a piecewise linear approximation in each time interval considering the control constraints

$$u_i(t) = \begin{cases} u_i^+, \text{if } \tilde{u}_i > u_i^+ \\ u_i^-, \text{if } \tilde{u}_i < u_i^- \\ \tilde{u}_i, \text{otherwise} \end{cases}, \tag{13}$$

where

$$\tilde{u}_i = \frac{q_{j+K(i-1)+1} - q_{j+K(i-1)}}{\Delta t}(t - (j-1)\Delta t) + q_{j+K(i-1)}, \tag{14}$$

$i = 1, \dots, m, \ j = 1, \dots, K.$

To solve the optimal control problem it is necessary to find the values of $(K+1)m$ parameters

$$\mathbf{q}_i = [q_1 \ldots q_{m(K+1)}]^T. \tag{15}$$

To find a solution an evolutionary algorithm is used.

All evolutionary algorithms include the following steps [12]:

- generation of possible set of solutions which is called a population;
- calculation of quality criterion value for each possible solution;
- variation or evolution of all or some subset of possible solutions;
- substitution of population elements: if a new possible solution is better than the old one before evolution, then substitution of the old possible solution with the new one.

One loop of all steps of evolutionary algorithm is a one generation. After several generations the best solution is determined and it is considered as a solution of the optimization problem. A number of generations and a number of elements in the initial population are the main parameters of all evolutionary algorithms.

Currently there are more than fifty evolutionary algorithms. Mostly they have exotic names, e.g. names after an animal or a nature phenomenon that inspired algorithm's creator. The most well-known of these algorithms is genetic algorithm [5,12]. It works with codes of possible solutions, so it can be used for searching for the optimal solution in a non-numerical space, where there is no metric norm for the distance between two possible solutions. This property of genetic algorithm is due to the evolution procedure, which carries crossover and mutation operators. These operators don't use arithmetic operations of summation, subtraction, multiplication or division.

All evolutionary algorithms differ by the third step - the evolution. The more information about the distribution of the quality criterion in the search space is used for evolution, the better evolutionary algorithm works for complex optimization problems. There are many evolutionary algorithms that work better than genetic algorithm in a metric search space, e.g. Particle Swarm Optimization algorithm (PSO) [12,13]. In this algorithm each possible solution is called particle. At particle's evolution step it uses information about the particle with current best value of quality criterion in the whole population, about the particle with best value of quality criterion among randomly selected particles and about own best value of quality criterion obtained in previous steps. The drawback of PSO algorithm and of many other evolutionary algorithms is the presence of a certain number of additional constant parameters that together with basic parameters must be configured for each specific optimization problem.

In this work we use Grey Wolf Optimizer algorithm (GWO) [12,14]. This algorithm uses only basic parameters of evolutionary algorithms, which are a number of generations and a number of elements in the population. Three solutions with the best values of quality criterion in current step are used for evolution of each possible solution. We modified this algorithm and made the number of chosen best current solutions used for evolution as an additional parameter of GWO algorithm.

GWO algorithm has the following steps:

1. Generate an initial population

$$Q = \{\mathbf{q}^1, \ldots, \mathbf{q}^H\}, \tag{16}$$

where

$$q^r_{i \cdot j} = \xi(q^+_i - q^-_i) + q^-_i, \tag{17}$$

$r = 1, \ldots, H, j = 1, \ldots, K+1, \xi$ is a random number from the interval $[0; 1]$, q^-_i and q^+_i are the minimum and the maximum values of parameters of the control function (13) respectively, $i = 1, \ldots, m$.

2. Calculate the value of quality criterion (11) for each possible solution

$$\mathbf{F} = \{f_1 = J_1(\mathbf{q}^1), \ldots, f_H = J_1(\mathbf{q}^H)\}, \tag{18}$$

where $J(\mathbf{q}^r)$ is the value of criterion (11) of the solution of system (1) with control function (13) obtained using vector $\mathbf{q}^r$.

Set the current value of generations iterator to $g = 1$.

3. Determine a set of N indices of the best possible solutions

$$\mathbf{I} = \{r_1, \ldots, r_N\}, \tag{19}$$

where

$$f_{r_1} \leq \cdots \leq f_{r_N} \leq \forall f_r, \ r \in \{1, \ldots, H\} \setminus \mathbf{I}. \tag{20}$$

4. Select a random possible solution from the population

$$r = \xi(H) \tag{21}$$

where $\xi(H)$ is a random integer number from 1 to H.

Calculate the following value

$$A = \sum_{i=1}^{N} \sum_{j=1}^{m(K+1)} q^{r_i}_j - L(2\xi - 1)|2\xi q^{r_i}_j - q^r_j|, \tag{22}$$

where

$$L = 2 - \frac{2g}{P}. \tag{23}$$

Perform an evolution of a possible solution $\mathbf{q}^r$

$$\tilde{q}^r_i = \frac{A}{N}, \ i = 1, \ldots, m(K+1). \tag{24}$$

Calculate the value of criterion (11) for a new vector $\tilde{\mathbf{q}}^r$. If $J_1(\tilde{\mathbf{q}}^r) < f_r$, then replace the old vector $\mathbf{q}^r$ with the new vector $\tilde{\mathbf{q}}^r$,

$$\mathbf{q} \leftarrow \tilde{\mathbf{q}}^r, \ f_r \leftarrow J_1(\tilde{\mathbf{q}}^r). \tag{25}$$

Repeat the step 4 $R \geq H$ times.

5. Increase the value of generations iterator

$$g \leftarrow g + 1 \tag{26}$$

and go to the step 3.

Repeat steps 3–5 P times.

In the end, after all iterations, the solution of the optimal control problem is a control function (13) obtained using the best solution among possible solutions in Q.

Our modified GWO algorithm has the following parameters: H is a number of possible solutions in a population; P is a number of generations; R is a number of evolutionary changes in one generation; N is a number of the best possible solutions used to perform an evolution. Our experience shows that GWO algorithm performs very well and highly suitable for solving the optimal control problem based on a direct approach.

4   Approximation of the Set of Extremals

As the optimal control problem was solved for each initial conditions from the set (7) we have all the received optimal controls and trajectories in the state space stored. The storage of discrete data was provided by determination of time interval $\Delta_s t$

$$D = \{D_1, \ldots, D_K\}, \tag{27}$$

where

$$D_i = \{(\tilde{\mathbf{u}}(0), \tilde{\mathbf{x}}^{0,i}), (\tilde{\mathbf{u}}(\Delta_s t), \tilde{\mathbf{x}}(\Delta_s t)), \ldots, (\tilde{\mathbf{u}}(W_i \Delta_s t), \tilde{\mathbf{x}}(W_i \Delta_s t))\}, \tag{28}$$

$$W_i = \left\lfloor \frac{t_f(\mathbf{x}^{0,i})}{\Delta_s t} \right\rfloor + 1, \tag{29}$$

$i = 1, \ldots, K$.

To solve the problem of the control system general synthesis it is necessary to find a control function in the form (5) that minimize quality criterion

$$J_2 = \sum_{i=1}^{K} \sum_{j=2}^{W_i} (\mathbf{x}((j-1)\Delta_s t, \mathbf{x}^{0,i}) - \tilde{\mathbf{x}}((j-1)\Delta_s t))^2 \rightarrow \min_{\mathbf{h}(\mathbf{x})}, \tag{30}$$

where $\mathbf{x}((j-1)\Delta_s t, \mathbf{x}^{0,i})$ is a solution of the system (6) at the moment $t = (j-1)\Delta_s t$ with initial condition $\mathbf{x}^{0,i}$, $j = 2, \ldots, W_i$, $i = 1, \ldots, K$. Here a starting point of calculation of the sum (29) is $j = 2$, since the control object's state at the moment $t = 0$ coincide for both the problem of the control system synthesis and the optimal control problem.

To solve the stated problem of the control system general synthesis, which is a search for a control function in the form (5) that minimize the quality criterion (30) we propose to use symbolic regression methods.

Symbolic regression methods search for a mathematical expression of a function in the form of specific code using genetic algorithm. The most popular method of symbolic regression is genetic programming (GP) [15]. GP encodes a mathematical expression in the form of computation tree. Elementary mathematical functions and arithmetic operations are located in nodes of the tree. Arguments of mathematical expression and constant parameters are located on leaves of the tree. The drawback of GP is that different mathematical expressions have different length of their codes. This is inconvenient for programming, since each code in a set of mathematical expression codes has a different length and its length can be changed after any of crossover operation. The second drawback of GP is that if an argument has to be written more than once in a mathematical expression, then it must also be located on the set of leaves several times.

There are many other symbolic regression methods that do not have these drawbacks. For example, in Cartesian genetic programming (CGP) [16] and in network operator (NOP) [2,4] a mathematical expression is represented in a constant-length code. This length does not change after crossover operations. In NOP a mathematical expression is represented in the form of a directed graph. Arguments are located at source nodes. Any argument appears in a source node only once, but this argument can appear several times in a mathematical expression. In CGP a mathematical expression is represented in the form of an integer matrix. Each column of the matrix is a call of an elementary function.

For comparison let us give an example of GP, CGP, and NOP codes for the mathematical expression

$$y = c_1 \sin(x_1) + x_1 \cos(c_2 x_1 + x_2). \tag{31}$$

A set of arguments for this expression is

$$F_0 = \{x_1, x_2, c_1, c_2\}. \tag{32}$$

A set of elementary functions is

$$F = \{f_1(z) = z, f_2(z) = \sin(z), f_3(z) = \cos(z), \\ f_4(z_1, z_2) = z_1 + z_2, f_5(z_1, z_2) = z_1 z_2\} \tag{33}$$

GP graph for the mathematical expression (31) is shown in Fig. 1. Here three inclusion of the argument x_1 in the mathematical expression required three leaves on the tree with this argument.

CGP code for the mathematical expression (31) is

$$Y = \begin{bmatrix} 2 & 5 & 5 & 4 & 3 & 5 & 4 \\ x_1 & c_1 & c_2 & 3 & 4 & x_1 & 6 \\ 0 & 1 & x_1 & x_2 & 0 & 5 & 2 \end{bmatrix}. \tag{34}$$

Here, the first line contains the numbers of functions. The second and the third lines contain arguments of functions. In the case when the second argument is not needed to call the function, zero is written. Arguments of functions are

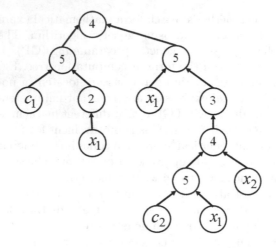

Fig. 1. Genetic programming graph for the mathematical expression

elements from the set (32) or the column number of this matrix, in which the function was already called.

NOP graph for the mathematical expression (31) is shown in Fig. 2. Arguments in this graph are located in source nodes, the numbers of functions of two arguments are located in remaining nodes. The numbers of functions of one argument are located on graph edges. The result of calculation is at the output node.

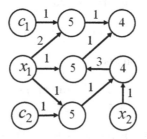

Fig. 2. Network operator graph for the mathematical expression

5   Computation Experiment

Consider the solution of the problem of the control system general synthesis for a mobile robot.

A mathematical model of the control object is given in the following form [17]

$$\dot{x}_1 = 0.5(u_1 + u_2)\cos(x_3),$$
$$\dot{x}_2 = 0.5(u_1 + u_2)\sin(x_3),$$
$$\dot{x}_3 = 0.5(u_1 - u_2),$$
$$(35)$$

where the control $\mathbf{u} = [u_1\ u_2]^T$ is constrained

$$u_i^- \leq u_i \leq u_i^+ : i = 1, 2. \tag{36}$$

$u_i^- = -10$, $u_i^+ = 10$, $i = 1, 2$.

Domain of initial conditions is

$$X_0 = \{x_i^- \leq x_i \leq x_i^+ : i = 1, 2, 3\}. \tag{37}$$

where $x_1^- = -2.5$, $x_2^- = -2.5$, $x_3^- = -5\pi/12$, $x_1^+ = 2.5$, $x_2^+ = 2.5$, $x_3^+ = 5\pi/12$.

Terminal state is

$$\mathbf{x}^f = [0\ 0\ 0]^T. \tag{38}$$

Quality criterion shows the time of reaching the terminal state (38)

$$J = \int_{x_1^-}^{x_1^+} \int_{x_2^-}^{x_2^+} \int_{x_3^-}^{x_3^+} \int_0^{t_f(\mathbf{x}^0)} dt = \int_{x_1^-}^{x_1^+} \int_{x_2^-}^{x_2^+} \int_{x_3^-}^{x_3^+} t_f(\mathbf{x}^0) \to \min_{\mathbf{u}=\mathbf{h}(\mathbf{x})}. \tag{39}$$

According to the proposed method, the optimal control problem should initially be solved for some set of initial conditions. Let us replace the domain (37) with the set of $M = 24$ initial conditions

$$\begin{aligned}
\tilde{X}_0 = \{&\mathbf{x}^{0,1} = [2.5\ 2.5\ 0]^T, \mathbf{x}^{0,2} = [0\ 2.5\ 0]^T,\\
&\mathbf{x}^{0,3} = [-2.5\ 2.5\ 0]^T, \mathbf{x}^{0,4} = [-2.5\ 0\ 0]^T,\\
&\mathbf{x}^{0,5} = [-2.5 - 2.5\ 0]^T\}.\mathbf{x}^{0,6} = [0 - 2.5\ 0]^T,\\
&\mathbf{x}^{0,7} = [2.5 - 2.5\ 0]^T, \mathbf{x}^{0,8} = [2.5\ 0\ 0]^T,\\
&\mathbf{x}^{0,9} = [2.5\ 2.5\ 5\pi/12]^T, \mathbf{x}^{0,10} = [0\ 2.5\ 5\pi/12]^T,\\
&\mathbf{x}^{0,11} = [-2.5\ 2.5\ 5\pi/12]^T, \mathbf{x}^{0,12} = [-2.5\ 0\ 5\pi/12]^T,\\
&\mathbf{x}^{0,13} = [-2.5 - 2.5\ 5\pi/12]^T, \mathbf{x}^{0,14} = [0 - 2.5\ 5\pi/12]^T,\\
&\mathbf{x}^{0,15} = [2.5 - 2.5\ 5\pi/12]^T, \mathbf{x}^{0,16} = [2.5\ 0\ 5\pi/12]^T\\
&\mathbf{x}^{0,17} = [2.5\ 2.5 - 5\pi/12]^T, \mathbf{x}^{0,18} = [0\ 2.5 - 5\pi/12]^T,\\
&\mathbf{x}^{0,19} = [-2.5\ 2.5 - 5\pi/12]^T, \mathbf{x}^{0,28} = [-2.5\ 0 - 5\pi/12]^T,\\
&\mathbf{x}^{0,21} = [-2.5 - 2.5 - 5\pi/12]^T, \mathbf{x}^{0,22} = [0 - 2.5 - 5\pi/12]^T,\\
&\mathbf{x}^{0,23} = [2.5 - 2.5 - 5\pi/12]^T, \mathbf{x}^{0,24} = [2.5\ 0 - 5\pi/12]^T\}.
\end{aligned} \tag{40}$$

For each initial condition from the set (40) the optimal control problem is solved using the following quality criterion

$$J_3 = t_f(\mathbf{x}^{0.i}) + \max\{|x_i^f - x_i(t_f(\mathbf{x}^{0.i}))| : i = 1, 2, 3\} \to \min_{\mathbf{u}(t)}, \tag{41}$$

where $t_f(\mathbf{x})^{0,j}$, $j = 1, \ldots, 24$, is defined in (9) considering $t^+ = 1s$, $\varepsilon_0 = 0.01$.

To solve the optimal control problem (35)–(41) the time interval is set to $\Delta t = 0.1$. Thus the dimension of a vector of parameters is $m(K + 1) = 20$, $\mathbf{q} = [q_1 \dots q_{20}]^T$.

The search for solutions of the optimal control problems was performed using GWO algorithm with the following parameters: $H = 512$, $P = 8192$, $R = 512$, $N = 8$.

Projections of found 24 optimal trajectories or in other words extremals onto the $\{x_1, x_2\}$ plane are shown in Fig. 3. Table 1 shows values of the quality criterion for each initial condition from the set (40).

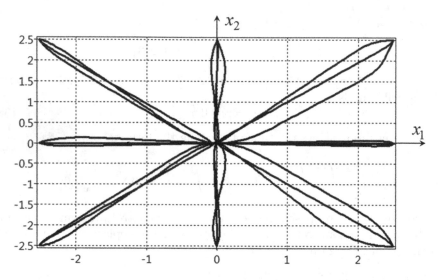

Fig. 3. Projections of found optimal trajectories onto the $\{x_1, x_2\}$ plane for 24 different initial conditions

Table 1. Optimal values of the quality criterion for different initial conditions.

Initial condition $\mathbf{x}^0$	J_3	Initial condition $\mathbf{x}^0$	J_3	Initial condition $\mathbf{x}^0$	J_3
$[2.5\ 2.5\ 0]^T$	0.53	$[0\ 2.5\ 0]^T$	0.57	$[-2.5\ 2.5\ 0]^T$	0.53
$[-2.5\ 0\ 0]^T$	0.25	$[-2.5\ -2.5\ 0]^T$	0.53	$[0\ -2.5\ 0]^T$	0.57
$[2.5\ -2.5\ 0]^T$	0.55	$[2.5\ 0\ 0]^T$	0.25	$[2.5\ 2.5\ 5\pi/12]^T$	0.5
$[0\ 2.5\ 5\pi/12]^T$	0.45	$[-2.5\ 2.5\ 5\pi/12]^T$	0.7	$[-2.5\ 0\ 5\pi/12]^T$	0.4
$[-2.5\ -2.5\ 5\pi/12]^T$	0.49	$[0\ -2.5\ 5\pi/12]^T$	0.45	$[2.5\ -2.5\ 5\pi/12]^T$	0.7
$[2.5\ 0\ 5\pi/12]^T$	0.39	$[2.5\ 2.5\ -5\pi/12]^T$	0.7	$[0\ 2.5\ -5\pi/12]^T$	0.46
$[-2.5\ 2.5\ -5\pi/12]^T$	0.49	$[-2.5\ 0\ -5\pi/12]^T$	0.39	$[-2.5\ -2.5\ -5\pi/12]^T$	0.7
$[0\ -2.5\ -5\pi/12]^T$	0.47	$[2.5\ -2.5\ -5\pi/12]^T$	0.5	$[0\ 0\ -5\pi/12]^T$	0.39

At the second step of the proposed method found optimal trajectories should be approximated by symbolic regression method. In the considered computational experiment we use Cartesian genetic programming. As mentioned above, CGP is free from drawbacks inherent in GP and is well suited for solving the considered problem.

The approximation of found extremals by CGP give us the following control function

$$u_i = \begin{cases} u_i^+, \text{if } \tilde{u}_i \geq u_i^+ \\ u_i^-, \text{if } \tilde{u}_i \leq u_i^+ \\ \tilde{u}_i, \text{otherwise} \end{cases}, \tag{42}$$

where

$$\tilde{u}_1 = 3q_1(x_1^f - x_1) + 2x_3^2 + 8q_3 \arctan(Q) + 4q_3(x_3^f - x_3), \tag{43}$$

$$\tilde{u}_2 = 6q_1(x_1^f - x_1) + 6x_3^2 - 8q_3 \arctan(Q) - 4q_3(x_3^f - x_3), \tag{44}$$

$Q = \frac{q_2(x_2^f - x_2)}{q_1(x_1^f - x_1)}$, $q_1 = 6.8261$, $q_2 = 10.92188$, $q_3 = 15.99707$, $q_4 = 9.85840$.

Figure 4 shows projections onto the plane $\{x_1, x_2\}$ of trajectories obtained by the object movement using the stabilization system (42)–(44) from eight initial conditions $[2.5\ 2.5\ 0]^T$, $[0\ 2.5\ 0]^T$, $[-2.5\ 2.5\ 0]^T$, $[-2.5\ 0\ 0]^T$, $[-2.5\ -2.5\ 0]^T$, $[0\ -2.5\ 0]^T$, $[2.5\ -2.5\ 0]^T$, $[2.5\ 0\ 0]^T$ (black lines) and points in the training set (red dots).

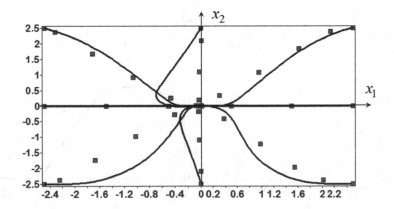

Fig. 4. Projections onto the plane $\{x_1, x_2\}$ of 8 trajectories obtained by the object movement using found stabilization system

It can be seen in Fig. 4 that some extremals were approximated with insufficient accuracy. It is because the trajectories on which the object has to rotate 90° are the most difficult for the approximation.

To confirm the results let us consider a solution which was obtained by the object movement using the stabilization system from the initial state that was not included in the training set of extremals, specifically

$$\mathbf{x}(0) = [1.25\ 2.5\ \pi/4]^T. \tag{45}$$

For this problem of moving the control object from the initial state (45) to the terminal state (38) there were obtained and compared solutions using the stabilization system (42)–(44) and using a direct approach to solve the optimal control problem by evolutionary GWO algorithm.

Graphs of found solutions are showed in Fig. 5. The graphs of solution obtained using the stabilization system (42)–(44) are showed in black, the graphs of solution obtained using a direct approach to solve the optimal control problem by evolutionary GWO algorithm are showed in red. Figure 6 shows the graphs of control values.

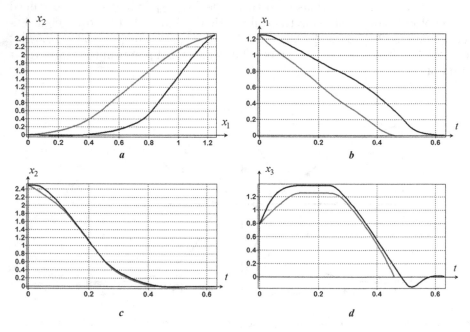

Fig. 5. Graphs of the control object movement from the initial state $[1.25\ 2.5\ \pi/4]^T$ to the terminal state $[0\ 0\ 0]^T$ obtained using the stabilization system found with proposed method (black) and using a direct approach to solve the optimal control problem by evolutionary GWO algorithm (red), $\boldsymbol{a}$ – movement trajectory projection onto the plane $\{x_1, x_2\}$; $\boldsymbol{b}$ – the graph $x_1(t)$; $\boldsymbol{c}$ – the graph $x_2(t)$; $\boldsymbol{d}$ – the graph $x_3(t)$

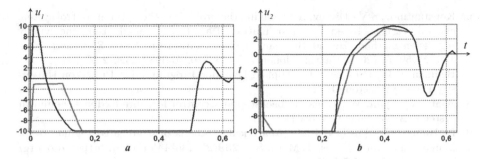

Fig. 6. Graphs of the control values obtained using the stabilization system found with proposed method (black) and using a direct approach to solve the optimal control problem by evolutionary GWO algorithm (red), $a - u_1(t)$; $b - u_2(t)$

6   Conclusion

A solution to the problem of the control system general synthesis using symbolic regression methods was considered. To ensure the proximity of found solution to the optimal one, a control function was obtained by approximating a set of optimal trajectories (extremals). Approximation was carried out with Cartesian genetic programming. A computational experiment of the control system synthesis using an approximation of the optimal trajectories set for a two-tracked mobile robot was conducted. In the approximation step a set of twenty four optimal trajectories obtained with grey wolf optimizer evolutionary algorithm was used. Computational experiment showed that the control function found allows one to obtain close to optimal control for the initial conditions that were not considered during the approximation of optimal trajectories. The maximum discrepancy of the quality criterion for the found solution is no more then 20% from the optimal one.

References

1. Boltyansky, V.G.: Mathematical Methods of Optimal Control. Holt, Rinehart and Winston, New York (1971)
2. Diveev, A.I., Sofronova, E.A.: Numerical method of network operator for multiobjective synthesis of optimal control system. In: 2009 IEEE International Conference on Control and Automation, pp. 701–708 (2009). https://doi.org/10.1109/ICCA.2009.5410619
3. Derner, E., Kubalík, J., Ancona, N., Babuška, R.: Symbolic Regression for Constructing Analytic Models in Reinforcement Learning. ArXiv, abs/1903.11483 (2019)
4. Diveev, A.I.: A numerical method for network operator for synthesis of a control system with uncertain initial values. J. Comput. Syst. Sci. Int. **51**(2), 228–243 (2012). https://doi.org/10.1134/S1064230712010066
5. Goldberg, D.E.: Genetic Algorithms in Search. Optimization and Machine Learning. Addison-Wesley, Boston (1989)

6. Konstantinov, S.V., Diveev, A.I.: Solving the problem of the optimal control system general synthesis based on approximation of a set of extremals using the symbol regression method. Herald Bauman Moscow State Tech. Univ. Ser. Instr. Eng. **131**(2), 59–74 (2020). https://doi.org/10.18698/0236-3933-2020-2-59-74
7. Bellman, R.E., Dreyfus, S.E.: Applied Dynamic Programming. Princeton University Press, Princeton (1971)
8. Zhang, S., Qian, W.: Dynamic backstepping control for pure-feedback nonlinear systems. ArXiv, abs/1706.08641 (2017)
9. Podvalny, S.L., Vasiljev, E.M.: Analytical synthesis of aggregated regulators for unmanned aerial vehicles. J. Math. Sci. **239**(2), 135–145 (2019). https://doi.org/10.1007/s10958-019-04295-w
10. Konstantinov, S.V., Diveev, A.I., Balandina, G.I., Baryshnikov, A.A.: Comparative research of random search algorithms and evolutionary algorithms for the optimal control problem of the mobile robot. Procedia Comput. Sci. **150**, 462–470 (2019). https://doi.org/10.1016/j.procs.2019.02.080
11. Grachev, I.I., Evtushenko, Y.G.: A library of programs for solving optimal control problems. USSR Comput. Math. Math. Phys. **19**(2), 99–119 (1980). https://doi.org/10.1016/0041-5553(79)90009-0
12. Diveev, A.I., Konstantinov, S.V.: Study of the practical convergence of evolutionary algorithms for the optimal program control of a wheeled robot. J. Comput. Syst. Sci. Int. **57**(4), 561–580 (2018). https://doi.org/10.1134/S106423071804007X
13. Kennedy, J., Eberhart, R.: Particle swarm optimization. In: Proceedings of IEEE International Conference on Neural Networks, ICNN 1995, vol. 4, pp. 1942–1948 (1995). https://doi.org/10.1109/ICNN.1995.488968
14. Mirjalili, S., Mirjalili, S.M., Lewis, A.: Grey wolf optimizer. Adv. Eng. Softw. **69**, 46–61 (2014). https://doi.org/10.1016/j.advengsoft.2013.12.007
15. Koza, J.R., Keane, M.A., Streeter, M.J., Mydlowec, W., Yu, J., Lanza, G.: Genetic Programming IV. Routine Human-Competitive Machine Intelligence. Springer, Boston (2003)
16. Miller, J.F.: Cartesian Genetic Programming. Springer, Heidelberg (2003)
17. Šuster, P., Jadlovska, A.: Tracking trajectory of the mobile Robot Khepera II using approaches of artificial intelligence. Acta Electrotechnica et Informatica **11**(1), 38–43 (2011). https://doi.org/10.2478/v10198-011-0006-y

Multi-point Stabilization Approach to the Optimal Control Problem with Uncertainties

Askhat Diveev$^{(\boxtimes)}$ ⓘ and Elizaveta Shmalko ⓘ

Federal Research Center "Computer Science and Control"
of the Russian Academy of Sciences, Moscow, Russia
aidiveev@mail.ru, e.shmalko@gmail.com

Abstract. The article is studying the solution of the optimal control problem on the implementation. The problem here is that to implement the received optimal control a stabilization system is needed. But construction of such system makes changes to the mathematical model of the control object, so that the received control is not more optimal for that object. The paper introduces the concept of feasibility of control systems. An approach based on multi-point stabilization to receive feasible solutions of the optimal control problem is proposed. According to the approach, stabilization system synthesis is solved firstly. The synthesis problem can be solved by any known analytical or technical method. In the paper a symbolic regression method is used for the numerical solution of the synthesis problem. As far as the solution of synthesis problem is received, the differential equations of the model as one-parameter mappings acquire the contraction property, which reduces small model uncertainties. Then positions of stable points in the state space are found such that when switching from one point to another in some time interval the control object will move from initial conditions to terminal one with an optimal value of a quality criterion. In the computational experiment the efficiency of the proposed approach comparative to the direct approach is shown in the presence of uncertainties.

Keywords: Optimal control · Phase constraints · Control synthesis · Contraction property · Feasibility principle · Genetic programming

1 Introduction

Solving control problems from a mathematical point of view is a qualitative change in the right sides of the system of differential equations describing the

The theoretical parts of work, Sections 1–4, were performed at support from the Russian Science Foundation (project No 19-11-00258). Experimental and computational part of the work, Sections 5, was performed at support from the Russian Foundation for Basic Research (project No 18-29-03061-mk).

© Springer Nature Switzerland AG 2020
N. Olenev et al. (Eds.): OPTIMA 2020, CCIS 1340, pp. 129–142, 2020.
https://doi.org/10.1007/978-3-030-65739-0_10

mathematical model of the control object. These changes are defined by a free control vector and a compact set that determines restrictions on the control.

Different control functions can provide different properties of control system or more exactly different properties of the system of differential equations. For example, when we solve the optimal control problem and find the control function as a function of time, we want that the system of differential equations of the mathematical model of the control object has a partial solution that from the given initial conditions hits the given terminal condition with the optimal value of the quality criterion. This found solution, in fact, cannot be directly implemented in practice. In real control systems to provide movement of the object on the found optimal trajectory it is necessary to build a stabilization system. But construction of the stabilization system changes the mathematical model of the object and the received control might be not optimal for the new model.

From the mathematical point of view the synthesis of stabilization system is an attempt to give an attractor property to the found optimal trajectory. As a result, in the problem of optimal control, it is necessary not only to formulate a requirement for solving the system of differential equations so that to achieve the terminal state with an optimal value of the quality criterion, but also to formulate additional requirements for the properties of this solution. The optimal solution must be Lyapunov stable or have the attractor property [1]. It can be that the attractor property for the optimal solution of the system of differential equations is redundant, and other weaker requirements are needed to implement the resulting solution.

This paper is devoted to the study of feasibility property of the solution of the optimal control problem. Based on a qualitative analysis [2] of the solutions of systems of differential equations, the concept of the feasibility of control systems is introduced. Feasibility is a property of control in which small changes in the model do not lead to a loss of quality. It is shown that the feasibility criterion was not initially laid down in the formulation of the optimal control problem.

Following the principle of feasibility, it is proposed to solve the problem of optimal control through the synthesis of a stabilization system. The stage of synthesis of the stabilization system allows to embed the control in the object so that the system of differential equations would have the necessary property of feasibility. In this case, the equilibrium point can be changed after some time, but the object maintains equilibrium at every moment of time.

A computational experiment is presented in which the sensitivity to disturbances of two controls obtained on the basis of the proposed approach and by a direct method is compared.

2   Feasibility Principle

Hypothesis A. *A mathematical model is feasible, if its errors do not increase in time.*

Definition 1. *The system of differential equations is practically feasible, if this system as a one-parametric mapping obtains a contraction property in the implementation domain.*

Consider a system of differential equations

$$\dot{\mathbf{x}} = \mathbf{f}(\mathbf{x}), \tag{1}$$

where $\mathbf{x} \in \mathbb{R}^n$.

Any ordinary differential equation is a recurrent description of a time function. A solution of the differential equation is a transformation from a recurrent form to a usual time function.

Computer calculation of the differential Eq. (1) has a form

$$\mathbf{x}(t + \Delta t) = \mathbf{x}(t) + \Delta t \mathbf{f}(\mathbf{x}(t)), \tag{2}$$

where t is an independent parameter, Δt is a constant parameter, and it is called a step of integration.

The right side of the Eq. (2) is a one-parametric mapping from space $\mathbb{R}^n$ to itself

$$F(\mathbf{x}, t) = \mathbf{x}(t) + \Delta t \mathbf{f}(\mathbf{x}(t)) : \ \mathbb{R}^n \to \mathbb{R}^n. \tag{3}$$

Let a compact domain D be set in the space $\mathbb{R}^n$. All solutions of the differential Eqs. (1), that are of our interest, belong to this domain. Therefore, for the differential Eqs. (1) the initial and terminal conditions belong to this domain

$$\mathbf{x}(0) \in D \subseteq \mathbb{R}^n, \ \mathbf{x}(t_f) \in D \subseteq \mathbb{R}^n, \tag{4}$$

where $\mathbf{x}(t_f)$ is a terminal point of the solution (1).

Theorem 1. *Let in the domain D for the mapping (3) the following property is performed*

$$\rho(\mathbf{x}^a(t), \mathbf{x}^b(t)) \leq \rho(F(\mathbf{x}^a(t), t), F(\mathbf{x}^b(t), t)), \tag{5}$$

where $\mathbf{x}^a(t) \in D$, $\mathbf{x}^b(t) \in D$, $\rho(\mathbf{x}^a, \mathbf{x}^b)$ is a distance between two points in the space $\mathbb{R}^n$

$$\rho(\mathbf{x}^a, \mathbf{x}^b) = \left\| \mathbf{x}^a - \mathbf{x}^b \right\|. \tag{6}$$

Then the mathematical model (1) is feasible if the domain $D \subseteq \mathbb{R}^n$ according to the hypothesis.

Proof. Let $\mathbf{x}(t) \in D$ is a known state of the system in the moment t and $\mathbf{y}(t) \in D$ is a real state of the system in the same moment. The error of the state is

$$\delta(t) = \rho(\mathbf{x}(t), \mathbf{y}(t)). \tag{7}$$

According to the mapping (3)

$$\delta(t + \Delta t) = \rho(F(\mathbf{x}(t), t), F(\mathbf{y}(t))). \tag{8}$$

And according to the condition (5) of the theorem

$$\delta(t) \leq \delta(t + \Delta t). \tag{9}$$

This proves the theorem $\square$

The condition (5) shows that the system of differential equations as a one-parametric mapping has contraction property.

Assume that the system (1) in the neighborhood of the domain D has one stable equilibrium point, and there is no other equilibrium point in this neighborhood

$$\mathbf{f}(\tilde{\mathbf{x}}) = 0, \tag{10}$$

$$\det(\lambda\mathbf{E} - \mathbf{A}(\tilde{\mathbf{x}})) = \lambda^n + a_{n-1}\lambda^{n-1} + \ldots + a_1\lambda + a_0 = \prod_{j=1}^{n}(\lambda - \lambda_j) = 0, \tag{11}$$

where $\mathbf{E}$ is a unit $n \times n$ matrix,

$$\mathbf{A}(\tilde{\mathbf{x}}) = \frac{\partial\tilde{\mathbf{f}}(\mathbf{x})}{\partial\mathbf{x}}, \tag{12}$$

$$\lambda_j = \alpha_j + i\beta_j, \tag{13}$$

$\alpha_j < 0$, $i = \sqrt{-1}$, $j = 1, \ldots, n$.

Theorem 2. *If for the system (1) there is a domain D that includes one stable equilibrium point (10)–(13), then the system (1) is practically feasible.*

Proof. According to the Lyapunov's stability theorem on the first approximation the trivial solution of the differential equation (1)

$$\mathbf{x}(t) = \tilde{\mathbf{x}} = \text{ constant} \tag{14}$$

is stable. This means, that, if any solution begins from other initial point $\mathbf{x}^0 \neq \tilde{\mathbf{x}}$, then it will be approximated to the stable solution asymptotically

$$\rho(\mathbf{x}(t + \Delta t, \mathbf{x}^a), \tilde{\mathbf{x}}) \leq \rho(\mathbf{x}(t, \mathbf{x}^a), \tilde{\mathbf{x}}), \tag{15}$$

where $\mathbf{x}(t, \mathbf{x}^a)$ is a solution of the differential equation (1) from initial point $\mathbf{x}^a$.

The same is true for another initial condition $\mathbf{x}^b$

$$\rho(\mathbf{x}(t + \Delta t, \mathbf{x}^b), \tilde{\mathbf{x}}) \leq \rho(\mathbf{x}(t, \mathbf{x}^b), \tilde{\mathbf{x}}). \tag{16}$$

From here follows, that the domain D has a fixed point $\tilde{\mathbf{x}}$ of contraction mapping [2], therefore distance between solutions $\mathbf{x}(t, \mathbf{x}^a)$ and $\mathbf{x}(t, \mathbf{x}^b)$ also tends to zero or

$$\rho(\mathbf{x}(t + \Delta t, \mathbf{x}^a), \mathbf{x}(t + \Delta t, \mathbf{x}^b)) \leq \rho(\mathbf{x}(t, \mathbf{x}^a), \mathbf{x}(t, \mathbf{x}^b)). \tag{17}$$

□

3 The Synthesized Optimal Control Problem Statement with Feasibility Conditions

It is given a mathematical model of the control object

$$\dot{\mathbf{x}} = \mathbf{f}(\mathbf{x}, \mathbf{u}), \tag{18}$$

where $\mathbf{x} \in \mathbb{R}^n$, $\mathbf{u} \in U \subseteq \mathbb{R}^m$, U is a compact set, $m \leq n$.

Initial conditions are given

$$\mathbf{x}(0) = \mathbf{x}^0 \in \mathbb{R}^n, \tag{19}$$

The terminal conditions are given

$$\mathbf{x}(t_f) = \mathbf{x}^f \in \mathbb{R}^n, \tag{20}$$

where t_f is a time of hitting the terminal conditions, t_f is not given, but limited

$$t_f \leq t^+, \tag{21}$$

t^+ is given.

The quality criterion is given

$$J = \int_0^{t_f} f_0(\mathbf{x}(t), \mathbf{u}(t)) dt \to \min_{\mathbf{u} \in U}, \tag{22}$$

It is necessary initially to find the control function in the form

$$\mathbf{u} = \mathbf{h}(\mathbf{x}^* - \mathbf{x}), \tag{23}$$

where $\mathbf{x}^*$ is a constant point in the state space, $\mathbf{x}^* \in \mathbb{R}^n$.

The control function (23) has the following properties:

A) There is a domain $D \subseteq \mathbb{R}^n$, where the control function (23) satisfies the restrictions on control

$$\forall \mathbf{x}^* \in D \to \mathbf{h}(\mathbf{x}^* - \mathbf{x}) \in U. \tag{24}$$

B) The differential equation system

$$\dot{\mathbf{x}} = \mathbf{f}(\mathbf{x}, \mathbf{h}(\mathbf{x}^* - \mathbf{x})) \tag{25}$$

has an equilibrium point $\tilde{\mathbf{x}}(\mathbf{x}^*)$ with the properties (10)–(13), therefore the system (25) is feasible according to definition 1.

At the second stage of the synthesized optimal control method it is necessary to find the function

$$\mathbf{x}^* = \mathbf{v}(t), \tag{26}$$

which provides to the partial solution of the system (25) the property of achievement of the terminal conditions (20) from the initial conditions (19) with the optimal value of the quality criterion (22).

Note, that at the second stage the searched function (26) has a dimension the same as the state space, and at the numerical solution the searched function (26) can be searched as a piece-constant one.

$$\mathbf{v}(t) = \mathbf{x}^{*,i}, \ \text{if} \ (i-1)\Delta \le t < i\Delta, \tag{27}$$

where $\mathbf{x}^{*,i}$ are found optimal values of point coordinates in the domain D, $i = 1, \ldots, K$, Δ is a given time interval,

$$K = \left\lfloor \frac{t^+}{\Delta} \right\rfloor. \tag{28}$$

The most difficult part of the problem statement (18)–(28) is to search the control function in the form (23). This sub-problem belongs to the class of the problems of control system general synthesis. It can be solved by any analytical, technical, or numerical methods. Analytical methods such as the analytic design of optimal controllers [3], the backstepping integrator [4,5] and the analytical construction of aggregated controllers [6,7] can be used for determined class of mathematical models (18) of control object. Engineers often use a technical approach to the synthesis when the control function is given, as a rule, intuitively, with the accuracy for values of parameters, that are searched by an optimization algorithm.

Now there are numerical methods of the control system synthesis that allow to find a mathematical expression for the control function. In recent years, symbolic regression methods have been used to numerically solve the control synthesis problem [8–15]. These methods search for a mathematical expression in the form of a special code with the help of an evolutionary genetic algorithm. Now more than ten symbolic regression methods are known. All these methods are differed in coding of mathematical expression and crossover and mutation operations of the genetic algorithm. In the present paper one of the latest symbolic regression method of variation Cartesian genetic programming is applied in the computational example section.

4   Variation Cartesian Genetic Programming

Variation Cartesian genetic programming (VCGP) is a Cartesian genetic programming with the application of the variation principle [16]. Therefore, let us consider initially Cartesian genetic programming [10].

To code a mathematical expression by a symbolic regression method it is necessary to set a set of elementary functions and to determine a set of arguments. Assume that elementary functions can have one, two or three arguments

$$F = \{f_1 = f_1(z), \ldots, f_l = f_l(z_1, z_2, z_3)\}. \tag{29}$$

Arguments of mathematical expression are variables and parameters

$$A = \{a_1 = x_1, \ldots, a_n = x_n, a_{n+1} = q_1, \ldots, a_{n+p} = q_p\}. \tag{30}$$

CGP-code of the mathematical expression is an integer matrix with four lines

$$\mathbf{C} = \begin{bmatrix} c_{1,1} & \cdots & c_{1,L} \\ \vdots & \vdots & \vdots \\ c_{4,1} & \cdots & c_{4,L} \end{bmatrix}, \tag{31}$$

where $c_{1,i}$ is the function number from the set of elementary functions (29),

$$c_{1,j} \in \{1, \ldots, l\}, \ j = 1, \ldots, L, \tag{32}$$

$c_{i,j}$ is the element number from the set of arguments (30) or $n + p$ plus the column number of the matrix (31) from 1 to $j - 1$,

$$c_{i,j} \in \{1, \ldots, n + p + j - 1\}, \ i = 2, 3, 4, \ j = 1, \ldots, L. \tag{33}$$

To decode the CGP-code an additional vector for storing the intermediate results of calculations is used

$$\mathbf{y} = [y_1 \ldots y_L]^T. \tag{34}$$

To calculate the result of the mathematical expression the following formula is applied

$$y_j = f_{c_{1,j}}(a_{c_{2,j}}, a_{c_{3,j}}, a_{c_{4,j}}), \ j = 1, \ldots, L. \tag{35}$$

If an elementary function has less than three arguments, then elements $c_{3,j}$ or/and $c_{4,j}$ aren't used. Total results of calculations are stored in the components of the vector (34).

Consider an example

$$u = q_1 x_1 + q_2 \exp(-x_2) \sin(q_3 x_3). \tag{36}$$

To code this mathematical expression the following sets are enough

$$F = \{f_1 = z, f_2 = -z, f_3 = \exp(z), f_4 = \sin(z), f_5 = z_1 + z_2, f_6 = z_1 z_2\}. \tag{37}$$

$$A = \{a_1 = x_1, a_2 = x_2, a_3 = x_3, a_4 = q_1, a_5 = q_2, a_6 = q_3\}. \tag{38}$$

Let the matrix of the code (31) uses ten columns, $L = 10$,

$$\mathbf{C} = [\mathbf{c}_1 \ldots \mathbf{c}_{10}]. \tag{39}$$

To code intermediate result, for example $q_1 x_1$, we use the number of elements from the set (37), (38)

$$\mathbf{c}_1 = \begin{bmatrix} 6 \\ 1 \\ 4 \\ 2 \end{bmatrix}. \tag{40}$$

Here the last element $c_{1,4} = 2$ isn't used.

If we want to use result $\mathbf{c}_1$ then the number $n + p + 1 = 3 + 3 + 1 = 7$ is used.

In total the following code is obtained

$$C = \begin{bmatrix} 6 & 2 & 3 & 6 & 6 & 4 & 6 & 5 & 1 & 1 \\ 1 & 2 & 8 & 5 & 6 & 11 & 10 & 13 & 14 & 15 \\ 4 & 3 & 4 & 9 & 3 & 7 & 12 & 7 & 1 & 2 \\ 2 & 4 & 5 & 7 & 8 & 8 & 9 & 1 & 3 & 5 \end{bmatrix}. \tag{41}$$

The last two columns have the same values, because for code of the mathematical expression (36) eight columns are enough, but we can't know in advance how many columns is needed.

The Variation Cartesian genetic programming uses the principle of small variations of the basic solution [16]. A small variation of CGP-code is a change of an element in the matrix (31). To write such small variations an integer vector with three components is used.

$$\mathbf{w} = [w_1 \ w_2 \ w_3]^T, \tag{42}$$

where w_1 is the line number of the matrix (31), $w_1 \in \{1, 2, 3, 4\}$, w_2 is the column number of the matrix $w_2 \in \{1, \ldots, L\}$, w_3 is a new value of the element c_{w_1, w_2}, if $w_1 = 1$, then $w_3 \in \{1, \ldots, l\}$, else $w_3 \in \{1, \ldots, n + p + w_2\}$.

For example, consider a small variation

$$\mathbf{w} = [2 \ 5 \ 9]^T, \tag{43}$$

of the code (41)

In the new code the element $c_{2,5} = 9$ instead of 6 or

$$\mathbf{w} \circ \mathbf{C} = \begin{bmatrix} 6 & 2 & 3 & 6 & 6 & 4 & 6 & 5 & 1 & 1 \\ 1 & 2 & 8 & 5 & 9 & 11 & 10 & 13 & 14 & 15 \\ 4 & 3 & 4 & 9 & 3 & 7 & 12 & 7 & 1 & 2 \\ 2 & 4 & 5 & 7 & 8 & 8 & 9 & 1 & 3 & 5 \end{bmatrix}. \tag{44}$$

The new code corresponds to the mathematical expression

$$\tilde{u} = q_1 x_1 + q_2 \exp(-x_2) \sin(\exp(-x_2)x_3). \tag{45}$$

Genetic operations of crossover and mutation are performed on the sets of variation vectors.

Let $\mathbf{W}_\alpha$ and $\mathbf{W}_\beta$ be two sets of variation vectors

$$\mathbf{W}_a = \{\mathbf{w}^{a,1}, \ldots, \mathbf{w}^{a,d}\}, \tag{46}$$

where $a = \alpha, \beta$, d is a given length of the sets.

Let k be a crossover point, $1 \le k \le d$. After crossover operation the following two new sets of small variation vectors are obtained

$$\mathbf{W}_\gamma = \{\mathbf{w}^{\alpha,1}, \ldots, \mathbf{w}^{\alpha,k-1}, \mathbf{w}^{\beta,k}, \ldots, \mathbf{w}^{\beta,d}\}, \tag{47}$$

$$\mathbf{W}_\sigma = \{\mathbf{w}^{\beta,1}, \ldots, \mathbf{w}^{\beta,k-1}, \mathbf{w}^{\alpha,k}, \ldots, \mathbf{w}^{\alpha,d}\}. \tag{48}$$

Mutation operation is a random generation of a new vector of small variations in the randomly selected position in the set of small variation vectors.

5   Computational Experiment

Consider the optimal control problem for the group of two mobile robots
The mathematical model of control object is

$$
\begin{aligned}
\dot{x}_1^j &= 0.5(u_1^j + u_2^j)\cos(x_3^j), \\
\dot{x}_2^j &= 0.5(u_1^j + u_2^j)\sin(x_3^j), \\
\dot{x}_3^j &= 0.5(u_1^j - u_2^j),
\end{aligned}
\tag{49}
$$

where $j = 1, 2$.

There are constraints on components of the control vectors

$$
-10 \le u_i^j \le 10,
\tag{50}
$$

where $j = 1, 2$, $i = 1, 2$.

The initial conditions are set

$$
\begin{aligned}
\mathbf{x}^1(0) &= \mathbf{x}^{0,1} = [0\ 0\ 0]^T, \\
\mathbf{x}^2(0) &= \mathbf{x}^{0,2} = [10\ 10\ 0]^T.
\end{aligned}
\tag{51}
$$

The terminal conditions are set

$$
\begin{aligned}
\mathbf{x}^1(t_f) &= \mathbf{x}^{f,1} = [10\ 10\ 0]^T, \\
\mathbf{x}^2(t_f) &= \mathbf{x}^{f,2} = [0\ 0\ 0]^T,
\end{aligned}
\tag{52}
$$

where

$$
t_f = \max\{t_{f,1}, t_{f,2}\},
\tag{53}
$$

$$
t_{f,i} = \begin{cases} t, & \text{if } \left\|\mathbf{x}^{f,i} - \mathbf{x}^i\right\| \le \varepsilon_0 = 0.01 \\ t^+ = 2.1, & \text{otherwise} \end{cases}.
\tag{54}
$$

The quality criterion is set

$$
J_1 = t_f + a_1 \sum_{k=1}^{2} \sum_{j=1}^{2} \int_0^{t_f} \vartheta\left(r_k - \sqrt{(x_1^j - x_{1,k})^2 + (x_2^j - x_{2,k})^2}\right) +
$$

$$
a_2 \int_0^{t_f} \vartheta\left(r_0 - \sqrt{(x_1^1 - x_1^2)^2 + (x_2^1 - x_2^2)^2}\right) +
$$

$$
a_3 \sum_{j=1}^{2} \sqrt{\sum_{i=1}^{3}(x_i^{f,j} - x_i^j)^2},
\tag{55}
$$

where $\vartheta(A)$ is Heaviside step function

$$
\vartheta(A) = \begin{cases} 1, & \text{if } A > 0 \\ 0, & \text{otherwise} \end{cases},
\tag{56}
$$

$r_1 = 3$, $r_2 = 3$, $x_{1,1} = 5$, $x_{1,2} = 5$, $x_{2,1} = 1$, $x_{2,2} = 9$, $r_0 = 2$, $a_1 = 3$, $a_2 = 3$, $a_3 = 2.5$.

At the first stage, the control system synthesis problem was solved by the Variation Cartesian genetic programming. The following control function was obtained

$$u_1^j = A_1 + B_1 + \text{sgn}(A_1)(\exp(|A_1|) - 1),$$
$$u_2^j = B_1 - A_1 - \text{sgn}(A_1)(\exp(|A_1|) - 1),$$

$$(57)$$

where

$$A_1 = q_1(x_3^{j,*} - x_3^j) + (x_2^{j,*} - x_2^j)\text{sgn}(x_2^{j,*} - x_2^j)\sqrt{|x_2^{j,*} - x_2^j|},$$

$$B_1 = 2(x_1^{j,*} - x_1^j) + q_2\text{sgn}(x_1^{j,*} - x_1^j),$$

$q_1 = 3.109$, $q_2 = 3.629$.

At the second stage, there were found points in the state space

$$\mathbf{x}^{j,*,i} = [x_1^{j,*,i} \ x_2^{j,*,i} \ x_3^{j,*,i}]^T,$$

$$(58)$$

where $j = 1, 2$, $i = 1, \ldots, K$.

Totally, three points $k = 3$ for each mobile robot were found.

The points in the space $\{x_1, x_2, x_3\}$ have the following coordinates: $\mathbf{x}^{1,*,1} = [4.0987 \quad 11.0967 \quad 0.05407]^T$, $\mathbf{x}^{2,*,1} = [5.3720 \quad 4.2932 \quad 0.4370]^T$, $\mathbf{x}^{1,*,2} = [6.6806 \quad 7,1450 \quad - 0.2273]^T$, $\mathbf{x}^{2,*,2} = [8.8362 \quad 1.7514 \quad 1.2929]^T$, $\mathbf{x}^{1,*,3} = [9.1204 \ 11.9757 \ 0,4329]^T$, $\mathbf{x}^{2,*,3} = [0.6469 \ - 1.0853 \ 0.3649]^T$.

To search for the points the particle swarm optimization (PSO) algorithm [17, 18] was used. Constraints for components of points were

$$-2 \le x_1^{j,*,i} \le 12,$$
$$-2 \le x_2^{j,*,i} \le 12,$$
$$-\pi/2 \le x_3^{j,*,i} \le \pi/2,$$

$$(59)$$

In the Fig. 1 the optimal trajectories on the plane $\{x_1, x_2\}$ for mobile robots are presented. Black lines are optimal trajectories of mobile robots, red circles are the phase constraints, small black squares are found points. Value of the functional (55) was 3.30865.

For comparative reasons this problem was solved as classical optimal control problem with the same functional and constraints. The optimal trajectories on the plane $\{x_1, x_2\}$ are presented in the Fig. 2. Value of the functional (55) was 2.1133.

To solve the optimal control problem the piece-wise linear approximation of control function was used. For this purpose axis t was cut on S intervals. In each interval i the control function had the following description

$$u_i^j = \begin{cases} u_i^+, & \text{if } \tilde{u}_i > u_i^+ \\ u_i^-, & \text{if } \tilde{u}_i < u_i^-, \ i = 1, 2, \\ \tilde{u}_i, & \text{otherwise} \end{cases}$$

$$(60)$$

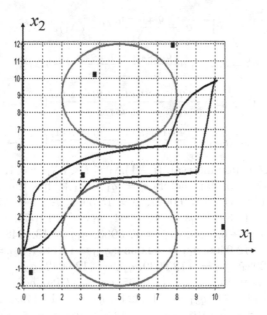

Fig. 1. Optimal trajectories of mobile robots on the plane $\{x_1, x_2\}$ with multi-point stabilization approach

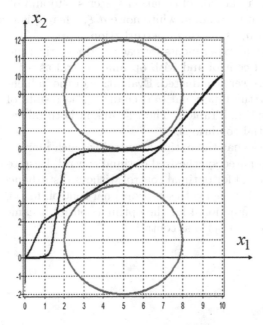

Fig. 2. Optimal trajectories of the classical optimal control problem

where

$$\tilde{u}_1^1 = q_i + \frac{q_{i+1} - q_i}{\Delta_1}(t - i\Delta_1), \tag{61}$$

$$\tilde{u}_2^1 = q_{i+S+1} + \frac{q_{i+S+2} - q_{i+S+1}}{\Delta_1}(t - i\Delta_1), \tag{62}$$

$$\tilde{u}_1^2 = q_{i+2(S+1)} + \frac{q_{i+1+2(S+1)} - q_{i+2(S+1)}}{\Delta_1}(t - i\Delta_1), \tag{63}$$

$$\tilde{u}_2^2 = q_{i+3(S+1)} + \frac{q_{i+1+3(S+1)} - q_{i+3(S+1)}}{\Delta_1}(t - i\Delta_1), \tag{64}$$

where $\Delta_1 = 0.21$ is a time interval, $(i-1)\Delta_1 \le t < i\Delta_1$, $i = 1, \ldots, S = 10$.

The optimal vector of parameters $\mathbf{q} = [q_1 \ldots q_{44}]^T$ has the following values: $q_1 = 9.0114$, $q_2 = 14.5534$, $q_3 = 9.0927$, $q_4 = 15.0562$, $q_5 = -1.6654$, $q_6 = 4.8062$, $q_7 = 18.9325$, $q_8 = 17.3911$, $q_9 = 10,0004$, $q_{10} = 19.1062$, $q_{11} = -12.1493$, $q_{12} = 10.2402$, $q_{13} = -6.4598$, $q_{14} = 17.3903$, $q_{15} = 13.1948$, $q_{16} = 19.7561$, $q_{17} = 13.5359$, $q_{18} = 12.9466$, $q_{19} = -2.0334$, $q_{20} = 17.1134$, $q_{21} = 10.6389$, $q_{22} = 18.6662$, $q_{23} = 10.2027$, $q_{24} = -12.8039$, $q_{25} = -12.6042$, $q_{26} = -11.5764$, $q_{27} = -17.5172$, $q_{28} = -13.9520$, $q_{29} = -11,4665$, $q_{30} = 5.6480$, $q_{31} = -14.6277$, $q_{32} = -13.9159$, $q_{33} = -17.9911$, $q_{34} = -14.4242$, $q_{35} = -14.4492$, $q_{36} = -18.1233$, $q_{37} = -3.2920$, $q_{38} = -15.8916$, $q_{39} = -13.8078$, $q_{40} = -13.1822$, $q_{41} = 0.7941$, $q_{42} = -16.1596$, $q_{43} = 13,8797$, $q_{44} = 14.7533$.

To check the feasibility property of the mathematical models of two robots the random disturbances were included. Two kind of disturbances were used. Disturbances of the right side of equation systems (49) and of initial conditions (51). Disturbances are a random white noise $a_r \xi$, where ξ is random value from -1 to 1, a constant a_r is a level of perturbation.

Ten computational experiments for each mathematical model (49) with the synthesized optimal control and with optimal control as a function of time for different noise levels were conducted. Evaluation of the influence of disturbances was determined by the functional (55) change. The results of the experiments are shown in Table 1 and Table 2. The last line in both tables shows results for disturbances of initial conditions.

The results show that the optimal control is much more sensitive to disturbances than the proposed synthesized optimal control. Despite the fact that the values of the functional for optimal control without disturbances were one and a half times less than for the synthesized optimal control, under disturbances the average value of the functional for the optimal control became two times worse than for the synthesized optimal control.

Table 1. Synthesized optimal control

No	Noise level	The best	Average	Standard deviation
1	0	3.0865	0	0
2	0.5	3.1275	3.4008	0.1969
3	1	3.1337	3.4100	0.2407
4	2	3.2503	4.2005	0.5108
5	0.1	3.1590	3.8847	0.5875

Table 2. Optimal control

No	Noise level	The best	Average	Standard deviation
1	0	2.1133	0	0
2	0.5	2.4302	6.4908	1.9916
3	1	7.2210	8.1248	1.0433
4	2	4.9516	8.4157	1.7849
5	0.1	4.6729	8.8921	1.7111

6   Conclusion

In the paper the definition of feasible control system is presented. The theorem that a contraction mapping posses the property of decreasing model errors is proved, therefore, all mathematical models with the stable equilibrium points in the state space are feasible. To achieve feasibility of the control system the synthesized optimal control method is proposed. In the work it is given the formulation of the optimal control problem, which includes the stage of synthesis of the stabilization system. After solution of the control synthesis problem the system of differential equations with found control function in the right side has always a stable equilibrium point in the state space. This means that the system of differential equations is a contraction mapping. Therefore this control system is feasible. On the second stage the optimal control problem is solved by searching for the stabilization points' positions. In the paper the new symbolic regression method of Variation Cartesian genetic programming is proposed for the solution of the control system synthesis problem. In the experimental part the optimal control problem of two robots is considered. To solve this problem two approaches were used, the synthesized optimal control and direct search for optimal control on the base of approximation of control by piece-wise linear functions. To search for positions of stabilization points in the state space and parameters of the piece-wise linear functions, the particle swarm optimization algorithm was used. The solution of the classical optimal control problem had better value of the functional than the solution by the synthesized optimal control method. After that both solutions were studied in conditions of disturbances. Experiments shown that the synthesized optimal control received solution that is more stable to disturbances, and this approach allows to receive practically feasible solutions.

References

1. Diveev, A.I., Shmalko, E.Y., Sofronova, E.A.: Multipoint numerical criterion for manifolds to guarantee attractor properties in the problem of synergetic control design. In: ITM Web of Conferences, vol. 18, p. 01001 (2018)
2. Kolmogorov, A.N., Fomin, S.V.: Elements of the Theory of Functions and Functional Analysis Metric and Normed Spaces. vol. 1, 130 p. Graylock Press. Rochester (1957)
3. Mizhidon, A.D.: On a problem of analytic design of an optimal controller. Autom. Remote Control **72**(11), 2315–2327 (2011). https://doi.org/10.1134/S0005117911110063
4. Krstic, M., Kanellakopoulos, M., Kokotovic, P.V.: Adaptive nonlinear control without overparametrization. Syst. Control Lett. **19**(3), 177–185 (1992)
5. Khalil, N.K.: Nonlinear Systems, 3rd edn. 750 p. Pearson Education Inc., Prentice Hall (1996) (2002). ISBN 0130673897
6. Podvalny, S.L., Vasiljev, E.M.: Analytical synthesis of aggregated regulators for unmanned aerial vehicles. J. Math. Sci. **239**(2), 135–145 (2019)
7. Kolesnikov, A.A., Kolesnikov, A.A., Kuz'menko, A.A.: Backstepping and ADAR method in the problems of synthesis of the nonlinear control systems. Mekhatronika, Avtomatizatsiya, Upravlenie **17**(7), 435–445. (2016). https://doi.org/10.17587/mau.17.435-445. (In Russian)
8. Koza, J.R.: Genetic Programming: On the Programming of Computers by Means of Natural Selection. MIT Press, Cambridge (1992)
9. O'Neil, M., Ryan, C.: Grammatical evolution. IEEE Trans. Evol. Comput. **5**, 349–358 (2001)
10. Miller, J.F.: Cartesian Genetic Programming, 344 p. The University of York Heslington. Springer (1998)
11. Zelinka, I., Oplatkova, Z., Nolle, L.: Analytic programming - symbolic regression by means of arbitrary evolutionary algorithms. IJSST **9**(9), 44–56 (2005)
12. Alibekov, E., Kubalik, J., Babuška, R.: Symbolic method for deriving policy in reinforcement learning. In: Proceedings of the 2016 IEEE 55th Conference on Decision and Control (CDC), 12–14 December, pp 2789–2799, Las Vegas (2016)
13. Diveev, A.I.: Numerical method for network operator for synthesis of a control system with uncertain initial values. J. Comput. Syst. Sci. Int. **51**(2), 228–243 (2012)
14. Diveev, A.I., Sofronova, E.A.: Numerical method of network operator for multiobjective synthesis of optimal control system. In: Proceedings of Seventh International Conference on Control and Automation (ICCA 2009), Christchurch, New Zealand, December 9–11, pp. 701–708 (2009)
15. Diveev, A.I.: Numerical methods for solution of the control synthesis problem. Moscow, RUDN University publishing, 192 p. (2019). (In Russian)
16. Diveev, A.I.: Small variations of basic solution method for non-numerical optimization. IFAC-PapersOnLine **48**(25), 28–33 (2015)
17. Kennedy, J., Eberhart, R.: Particle swarm optimization. In: Proceedings of IEEE International Conference on Neural Networks, vol. IV, pp. 1942–1948, Perth (1995)
18. Diveev, A.I., Konstantinov, S.V.: Study of the practical convergence of evolutionary algorithms for the optimal program control of a wheeled robot. J. Comput. Syst. Sci. Int. **57**(4), 561–580 (2018)

Optimization in Economy, Finance and Social Sciences

Sociality is a Mechanism for Collective Action Dilemma Resolution

Tatiana Babkina[1,5][✉] [iD], Anna Sedush[1], Olga Menshikova[1,2] [iD],
and Mikhail Myagkov[3,4] [iD]

[1] Moscow Institute of Physics and Technology (State University),
9 Institutskiy per., Dolgoprudny, Moscow Region 141701, Russian Federation
babkinats@yandex.ru
[2] The Russian Presidential Academy of National Economy and Public
Administration, 84 Vernadsky Prospekt, Moscow 119606, Russian Federation
[3] Tomsk State University, 36 Lenina Avenue, Tomsk 634050, Russian Federation
[4] University of Oregon, 1585 E. 13th Avenue, Eugene, OR 97403, USA
[5] Federal Research Center Computer Science and Control,
Russian Academy of Sciences, Moscow 119333, Russian Federation

Abstract. Numerous studies (mostly in economics) address the issue
of collective action dilemmas for public goods, but few focus on psychological factors that affect individual decisions, and can be instrumental
in designing effective mechanisms of public good provision. This paper
reports on a series of laboratory experiments where "human sociality" is
used as key variable in public goods type games. We show that once social
ties have been formed (even after short-term socialization) among group
members, it facilitates strategies that lead to much higher rates of public
goods provision, and, thus make collective action a success. Moreover,
the amount of participants who choose individual strategies decreases in
two times. That is, it solves the common problem of free-riders because
participants begin to exhibit more socially responsible character. We also
demonstrate that results hold in situations that involve risk, and females
tend to be better contributors to public goods than their male counter-
parts.

Keywords: Social dilemma · Cooperation · Sociality · Collective-risk
social dilemma · Free-riders

1 Introduction

Our world is arranged in such a way that everything is interconnected. Every
real decision has a response in the future, whether we like it or not. Therefore, even in ancient civilizations, thoughts arose that it is necessary to preserve
what we have [1,2]. However, examples of irrational use of resources from the
world history suggest that public goods problems are still relevant and have

This research was supported by The Tomsk State University competitiveness improvement program and by the grant in RFBR 19-01-00296A.

N. Olenev et al. (Eds.): OPTIMA 2020, CCIS 1340, pp. 145–157, 2020.
https://doi.org/10.1007/978-3-030-65739-0_11

not been fully studied yet [3–5]. Therefore, the question "How can we improve the use of public goods?" is still actual. Shifting from individual strategies to collective strategies is an important task for society. Is it possible? What mechanisms can be involved in this process? Previously, it was found that in the Prisoner's Dilemma game, introduction of social interaction between participants—socialization—raises and maintains the level of cooperation [6,7]. Therefore, we decided to check whether socialization would influence the behavior patterns and decision-making in social dilemmas. For this purpose, we chose the problem of the public goods using with the existence of collective risk.

The use of a modification of the commonly known public goods game [8–10] was described for the first time by Milinski et al. [11], where it was presented as collective-risk social dilemma [12]. The idea is that a group of individuals is first given a certain resource – a collective or public good, from which, at the first stage of the game, they can repeatedly take their own points. At the second stage of the game, players in group invest in the collective good gradually. Their goal is to reach the definite target sum, otherwise, they may lose all the points with a certain degree of probability, which is a collective risk. In the described study, the percentage of losses was 90%, 50%, and 10%. In the case of 90%, half the groups achieved the target sum; in the case of 50%, it was only one group; in the case of 10%, none of the groups. Tavoni et al. proposed to modify the game by artificially dividing participants into "rich" and "poor" at the first stage of the game [13]. This made it possible to prove that the "poor" cannot save the collective good from collapse, if the "rich" do not contribute to the common good in accordance with their available capital. Thus, it is reliably known that the behavior of participants in this game will be affected by the probability of loss and the initial economic situation of each participant prior to the stage of contribution. In a review paper [14], it is stated that a lot of various factors can affect players' decisions in the public goods game; however, the factors listed above are significant. Besides, it is assumed that participants contribute about 40–60% in a one-shot game or in the first period of finitely repeated game; then, the participants drastically reduce their contributions, but do not reduce them to zero; however, about 70% of participants do not contribute anything in the last round, if this information is available. Those who believe that the rest will cooperate are more likely to cooperate themselves. However, rational egoists will not contribute anything regardless of what is happening around them [15,16]. It postulates the free-riders problem [17]. According to free riding hypothesis, players should invest nothing in the collective good if the game is finite [18].

The next important factor affecting behavior in social dilemmas is gender. In experiments by Nowell and Tinkler [19], using the design as in Isaac and Walker [20,21], it was shown that all-female groups are more cooperative in the public goods game than all-male or mixed groups. Although in other social dilemmas, such as the Prisoner's Dilemma, women cooperate in some cases less than men [22,23]. In previous studies, we revealed that in groups of strangers, females cooperate better than males, whereas in a socialized group, on the contrary, males are more cooperative [24].

Since the dilemma is social, any information about other participants will be taken into account, and further actions will be adjusted accordingly [25]. In Frey and Meier [26], it was found that most players change their behavior to prosocial when they know that at least one of them contributed at the previous stages. However, some people have strictly fixed behavior patterns, i.e., they contribute either always or never. A similar result was obtained at a ski resort in Sweden [27]. The effect of the influence of other participants was described in detail also in Fischbacher et al. [28]. Their paper introduces the concept of conditional cooperators, viz., the tendency of people to increase their cooperativeness when others commence to cooperate more. In their experiment, about 50% of people acted as conditional cooperators, although a third of the total group were free riders. But there is no evidence that this behavior can be influenced by the other external factors such as sociality. Although, it was found that even a small amount of social interaction (socialization) in groups of strangers increases cooperation in the Prisoner's Dilemma and makes it comparable with cooperation within a group of friends [6, 7, 29]. So, participants after socialization have more a sustainable behavior regardless of the others' behavior. A similar study was performed by Keser and Van Winden [30] for the public goods game. It was found that partners have much higher cooperation than strangers. However, there are no confirmed facts related to the way behavior of participants in a collective-risk social dilemma will change if a group of strangers is socialized. Will we observe the changes in strategies after socialization? Will the behavior of males and females change in the same way?

We hypothesize that:

1. Socialization helps to reduce the number of the free-riders.
2. Socialized groups more often reach the required level of contribution than non-socialized ones.
3. Males and females have the different strategies in the collective risk social dilemmas games.

2   Materials and Methods

The experiments were conducted at the Laboratory of Experimental Economics of the Moscow Institute of Physics and Technology. Students of MIPT (N = 96, 62 males) took part in the experiment. Participants for the experiments were invited through advertisements in the group of laboratory on the social networking site https://vk.com/ee_phystech. In total, 8 experiments were performed during the spring semester of 2016. Twelve participants were involved in each experiment. Most of the students were initially unfamiliar with each other. For this purpose, the faculty, course, group number and native town of a participant were taken into account when recruiting participants. After the end of each treatment, participants provided feedback about the experiments, received payments and left the experimental facility. Tomsk State University Human Subjects Committee approved the study procedures involving human participants. Written informed consents were obtained from participants. To conduct the game, a specialized tool for designing and carrying out experiments in a group of experimental economics, z-Tree, developed at the University of Zurich, was used [31].

2.1 Experimental Procedure

Part 1. In each session the participants were randomly divided into two groups of 6 people each. Participants did not know with whom they were in the same group. Then, participants were invited to play a game in a computer. The description of the game will be presented below.

Part 2. The participants were asked to take part in an interaction, referred to as socialization [6, 32]. First, the participants remembered each other's names by means of the Snowball game, the rules of which are as follows: (1) all the participants sit down in a circle, and then the first one calls out his or her name and a personality characteristic, starting with the same letter as the name; the next participant repeats the name and characteristic of the first participant and states his or her own name and characteristic; (2) the game comes in chain order to the closing player in the circle, who states all the names and characteristics; (3) in the reverse order, the participants share personal information: native town, faculty, hobby, interests. Then, two captains were selected among the participants on a voluntary basis, and the rest of the participants must choose which team, i.e., which captain, they want to join. The participants were distributed among the teams according to the following algorithm: (1) the 2 captains remain in the audience, and the other 10 participants leave it; (2) the participants one by one randomly enter the audience and say which team they want to join and why; (3) the group is considered to be complete when it has 6 participants. At the end of the socialization stage, the participants were given 5 min to work in groups to find 5 common characteristics (eye color, favorite food or movie, etc.) and the name for the group. This is the end of the socialization stage.

Part 3. In each session the participants played the same game as in part 1, but they were already in the groups assembled in part 2. That is, in part 3 the participants knew exactly, with whom they were in the same group.

The game presented to the participants was divided into two phases:

In Phase 1, a group of P participants initially had a common fund (collective good) with X points. During N periods, participants had the opportunity to extract 0, 1, 2, 3 or 4 points from the common fund. After each periods participants saw how many points were extracted by the other subjects in the group. However, this procedure was anonymous. The following equation had to be fulfilled:

$$N * P * 4 = X \tag{1}$$

where N is the number of periods, P stands for the number of participants, and X means the initial sum of points in the common fund.

After Phase 1, each participant had a certain total profit, which was equal to the number of points he or she had extracted during N periods of Phase 1.

The total sum of points extracted by all the participants was also known (it was Y points).

In Phase 2, the participants had an opportunity to return (invest) points into the common fund using the points received after Phase 1.

During N periods, the participants could contribute 0, 1, 2, 3 or 4 points to the common fund. After each periods participants saw how many points were contributed by the other subjects in the group. However, this procedure was anonymous. The goal of a group of participants was to return (contribute) 53% of the total sum extracted in Phase 1 (Y) [33]. If the required level of contribution (0.53Y) was reached, then all the participants received the total number of points that they had by the end of Phase 2 (i.e., the difference between individual extraction and the participant's individual contribution into the common fund). If the group did not reach the required level of contribution (0.53Y), then it did not receive anything (with a certain degree of probability). The probability of losing everything – P_{LOSS} - was determined based on the percentage between Y and X:

If Y $\leq$ 25% of X, then $P_{LOSS} = 2/12$,
If Y $>$ 25% of X & $\leq$ 50% of X, then $P_{LOSS} = 6/12$,
If Y $>$ 50% of X & $\leq$ 75% of X, then $P_{LOSS} = 9/12$,
Finally, if Y $>$ 75% of X, then $P_{LOSS} = 11/12$.

P_{LOSS} were determined by drawing one card out of 12 cards, numbered from 1 to 12. If, as a result of drawing cards, the probability requirement was met, then all the participants received the total number of points left by the end of Phase 2. Otherwise, they lost all earned points.

In our experiments, each group consisted of P = 6 participants. The number of game periods in 4 experiments was N = 10 periods; in the remaining 4 experiments, N = 5 periods. Accordingly, in a 10-period game, the initial number of points in the common fund is X = 240 points, in a 5-period game, X = 120 points.

In the first 4 experiments with 10-period games, the participants played one game in part 1, and one game in part 3. In the next 4 experiments with a 5-period game, the participants played two games in a row in part 1 and two games in part 3. The number of games was doubled in order to detect the possible trainability effect.

2.2   Theoretical Results

Theoretically, it is to a participant's advantage to extract 4 points in each period in order to obtain all the points. As for Phase 2 (reverse contribution), the return rate of 53% was not chosen by chance. If participants had to return 50%, then the strategy would be obvious, viz., to return 2 points in each period. However, taking into account the rate of 53%, at least a few participants should donate in one of the periods more than 2 points to gain the necessary amount.

2.3   Free-Riders

In order to calculate the number of free riders, the required level of contribution per one subject is compared with the contribution of the subject. If a participant contributes less more than on one the required contribution level per subject, i.e. he or she decided to save money hoping on the other subjects, he or she is considered as a free-rider.

3   Results and Discussion

3.1   Strategies, Social and Gender Influence

First of all, it was calculated the amount of the unsuccessful games, i.e., in which participants could not reach the required level of contribution. In total, 24 games were played before socialization and 24 games after socialization. Before socialization, the percentage of unsuccessful games was 41.7% (10 games from 24). After socialization, the percentage of unsuccessful games was 8.4% (2 games from 24). Thus, participants started to choose the collective strategies rather than individual.

Before socialization, the prevalence of the individual strategies was the main reason of not reaching the required level of contribution. Participants tried to save their money (individual strategy) hoping that there was someone who contribute the missing part of the required amount (collective strategy). After socialization, participants in groups started to choose the similar strategies which led to success.

If we consider Standard Deviation (SD) for extractions and contributions, we can see that SD decreases after socialization. This can be interpreted as a smaller data spread; i.e., participants begin to act more cohesively.

The standard deviation of the average level of extractions before socialization is 0.74, but after socialization it equals to 0.27. Similarly, the standard deviation for the average contribution levels before socialization is 0.61; after socialization, 0.30 (Table 1). It supports the idea that participants started "to think" in the similar way.

Table 1. Average levels of extraction and contribution before and after socialization.

Game phases	Game before socialization		Game after socialization		Z	p-value
	M	SD	M	SD		
Extraction	3.50	0.74	3.92	0.27	−6.95	<0.0001
Contribution	1.85	0.61	2.15	0.30	−5.06	<0.0001

Moreover, participants started to choose more logical but collective strategies increasing the average levels of extraction ($Z = -6.95$, p < 0.001, Wilcoxon signed-rank test) and contribution (Z = −5.06, p < 0.001, Wilcoxon signed-rank test).

Hence, socialization not only improves the responsibility of participants, but also equalizes their strategies, making them consistent, which enables them to attain the required level of contribution.

It is interesting to notice that the difference between males and females exists only in the second games both before and after socialization in experiments with 5 periods. Before socialization in the second game, females contribute 2.19; and males, 1.76 (Table 2 and Table 3). This means that, statistically, females contribute much more than males (Wilcoxon rank-sum test: Z = −2.17, p-value = 0.03) (see Fig. 1).

Table 2. Average level of contribution among males in 5-period games.

Males' contribution	Game 1		Game 2		Z	p-value
	M	SD	M	SD		
Before socialization	1.60	0.70	1.76	0.62	−1.14	0.25
After socialization	2.04	0.38	2.13	0.32	−0.29	0.77

The table shows the difference between male's contribution in Game 1 and male's contribution Game 2 before and after socialization (Wilcoxon signed-rank test).

Table 3. Average level of contribution among females in 5-period games.

Females' contribution	Game 1		Game 2		Z	p-value
	M	SD	M	SD		
Before socialization	1.82	0.78	2.19	0.58	−2.4	0.02
After socialization	2.22	0.32	2.30	0.31	−1.3	0.2

The table shows the difference between female's contribution in Game 1 and female's contribution Game 2 before and after socialization (Wilcoxon signed-rank test).

After socialization in the second game, the contribution of females equals 2.30, which is significantly greater than the contribution of males, which equals 2.13 (Tables 2 and 3) (Wilcoxon rank-sum test: Z = −1.9, p-value = 0.05) (see Fig. 2).

This result corresponds to the study [19]. So, not only in the public goods game but also on in a collective risk social dilemma females are more cooperative.

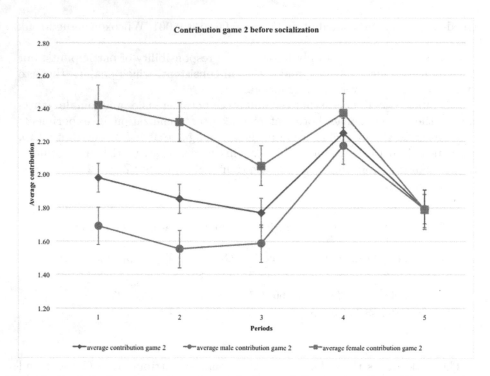

Fig. 1. Contributions in the second game before socialization for all participants, males and females.

3.2 Free-Riders

One of the most important indicator of the behavior in collective dilemmas is the amount of free-riders. Free-rider's existing in the group is the cause of loss everything in the game. Before socialization, it was 36% of the free-riders (38% in the 10-period games and 34% in the 5-period games). After socialization, the amount of free-riders decreased in two times and was equaled to 18% (21% in the 10-period games and 14% in the 5-period games). Actually, this result is strongly correlated with the result about increasing the level of successful games after socialization. However, it emphasizes the socialization influence on the decision making in collective dilemmas.

3.3 Learning Effect

Except for socialization, learning effect as well can influence the decision making in social dilemmas. To exclude or include this factor there were experiments, where subjects participated in the two similar games before and in the two similar games after socialization.

The results show that there is not statistically significant difference between contribution in game 1 and game 2 after socialization (Wilcoxon signed-rank

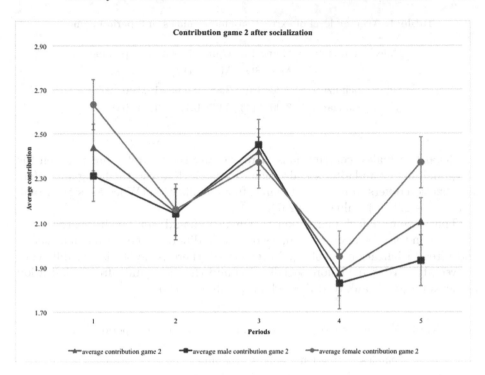

Fig. 2. Contributions in the second game after socialization for all participants, males and females.

test: Z = −0.96, p-value = 0.34). That is, the behavior during extraction phase before and after socialization between game 1 and game 2 is different as well as the behavior during contribution phase before socialization. It indicates that learning effect can influence the decision making. However, it was found the difference in learning effect between males and females.

Considering 5-period games, we can observe that, in part 1 before socialization, the behavior of male participants in two consecutive games coincides; however, it differs from the behavior of participants in part 3 after socialization.

Table 4 shows that there is no difference in males' extractions in the two games before socialization, (Wilcoxon signed-rank test: Z = −1.23, p-value = 0.22). Besides, no difference for males is registered between the two games after socialization (Wilcoxon signed-rank test: Z = −1.73, p-value = 0.083). However, a certain difference between the last game before socialization and the first game after socialization is observed (Wilcoxon signed-rank test: Z = −2,19, p-value = 0.03).

The table shows the difference between male's extraction in Game 1 and male's extraction Game 2 before and after socialization (Wilcoxon signed-rank test).

Table 4. Average level of extraction among males in 5-period games.

Males' extraction	Game 1		Game 2		Z	p-value
	M	SD	M	SD		
Before socialization	3.20	1.01	3.58	0.63	−1.23	0.22
After socialization	3.90	0.33	3.97	0.15	−1.73	0.08

Regarding males' contributions, here, we also see no difference between the two games before and after socialization (Table 2). However, the last game before socialization differs from the first game after socialization (Wilcoxon signed-rank test: $Z = -2.216$, p-value $= 0.0267$).

The learning effect in relation to females should not be disregarded. For instance, in the case of extraction, we observe a difference between games before and after socialization (Table 5). Nevertheless, there is no statistical difference between the last game before socialization and the first game after socialization (Wilcoxon signed-rank test: $Z = -1.14$, p-value $= 0.25$).

Table 5. Average level of extraction among females in 5-period games.

Females' extraction	Game 1		Game 2		Z	p-value
	M	SD	M	SD		
Before socialization	3.14	0.96	3.63	0.49	−2.97	0.003
After socialization	3.77	0.40	3.94	0.16	−2.23	0.025

The table shows the difference between female's extraction in Game 1 and female's extraction Game 2 before and after socialization (Wilcoxon signed-rank test).

The first two games before socialization differ in terms of females' contribution (Wilcoxon signed-rank test: $Z = -2.4$, p-value $= 0.02$), but games after socialization are not statistically distinguishable (Wilcoxon signed-rank test: $Z = -1.3$, p-value $= 0.2$) (Table 3). Furthermore, the statistical test does not reveal a difference in games before and after socialization (Wilcoxon signed-rank test: $Z = -0.14$, p-value $= 0.89$).

4   Conclusion

Thus, in this study we focused on investigating the behavior patterns of people in the case of using public goods with probability of collective risk under condition of social influence as well as gender differences in behavior. It has been revealed that socialized groups are more successful in games than non-socialized ones. After socialization, participants tend to make decisions more cohesively than before socialization. Implemented strategies are associated with organization and cooperation, which eventually leads a group to win, rather than to a critical situation. Also, the amount of free-riders drops in two times after socialization, that is participants started to choose collective strategies rather than individual.

In addition, based on the results of the experiments, it has been concluded that on average females are more compliant in the case of using the public good. For example, females sacrifice their personal gain in order to ensure success of the whole group more often than males. However, the learning factor affects females more strongly. In other words, retry of games leads to changes in strategies. On the contrary, males demonstrate static behavior in relation to retry of games under condition of unchanging social context.

Obviously, all these factors should be taken into consideration and server the purposes of public projects, strategies for the development of modern society, and promotion of the proper use of available resources [34–36].

References

1. Butzer, K.W.: Early Hydraulic Civilization in Egypt. University of Chicago Press, Chicago (1976)
2. Hughes, J.D.: Ecology in Ancient Civilizations. University New Mexico Press, Albuquerque (1975)
3. De Liddo, A., Concilio, G.: Making decision in open communities: collective actions in the public realm. Group Decis. Negot. **26**(5), 847–856 (2017). https://doi.org/10.1007/s10726-017-9543-9
4. Hardin, G.: The tragedy of the commons. Science **162**, 1243–1248 (1968). https://doi.org/10.1126/science.162.3859.1243
5. Ostrom, E.: Governing the Commons. Cambridge University Press, Cambridge (2015)
6. Babkina, T., Myagkov, M., Lukinova, E., Peshkovskaya, A., Menshikova, O., Berkman, E.T.: Choice of the group increases intra-cooperation. In: CEUR Workshop Proceeding, vol. 1627, pp. 13–23 (2016)
7. Berkman, E.T., Lukinova, E., Menshikov, I., Myagkov, M.: Sociality as a natural mechanism of public goods provision. PLoS ONE (2015). https://doi.org/10.1371/journal.pone.0119685
8. Dehez, P.: Cooperative provision of indivisible public goods. Theor. Decis. **74**(1), 13–29 (2013). https://doi.org/10.1007/s11238-012-9311-x
9. Ledyard, J.O.: Public goods: a survey of experimental research. In: Kagel, J.H., Roth, A.E. (eds.) Handbook of Experimental Economics, pp. 111–194. Princeton University Press, Princeton (1995)
10. Lichbach, M.I.: The repeated public goods game: a solution using tit-for-tat and the Lindahl point. Theor. Decis. **32**(2), 133–146 (1992). https://doi.org/10.1007/BF00134048
11. Milinski, M., Sommerfeld, R.D., Krambeck, H.-J., Reed, F.A., Marotzke, J.: The collective-risk social dilemma and the prevention of simulated dangerous climate change. Proc. Nat. Acad. Sci. **105**(7), 2291–2294 (2008)
12. Kline, R., Mahajan, A., Tingley, D.: Distributional equity in climate change policy: responsibility, capacity, and vulnerability (2016). https://wp.nyu.edu/cesspolicon2016/wp-content/uploads/sites/3319/2016/04/distributionequity.pdf. Accessed 20 June 2020

13. Tavoni, A., Dannenberg, A., Kallis, G., Loschel, A.: Inequality, communication, and the avoidance of disastrous climate change in a public goods game. Proc. Nat. Acad. Sci. **108**(29), 11825–11829 (2011)
14. Ostrom, E.: Collective action and the evolution of social norms. J. Nat. Resour. Policy Res. **6**(4), 235–252 (2014)
15. Goldfarb, R.S., Griffith, W.B.: Amending the economist's "rational egoist-model to include moral values and norms, part 2: alternative solutions'. Soc. Norms Econ. Inst. **2**, 59–84 (1991)
16. Jamieson, D.: Rational egoism and animal rights. Environ. Ethics (1981). https://doi.org/10.5840/enviroethics19813244
17. Andreoni, J.: Why free ride?: strategies and learning in public goods experiments. J. Public Econ. **37**(3), 291–304 (1988)
18. Friedman, J.W.: Game Theory with Applications to Economics. Oxford University Press, New York (1986)
19. Nowell, C., Tinkler, S.: The influence of gender on the provision of a public good. J. Econ. Behav. Organ. **25**(1), 25–36 (1994)
20. Isaac, R.M., Walker, J.M.: Communication and free-riding behavior: the voluntary contribution mechanism. Econ. Inq. **26**(4), 585–608 (1988a)
21. Isaac, R.M., Walker, J.M.: Group size effects in public goods provision: the voluntary contributions mechanism. Q. J. Econ. **103**(1), 179–199 (1988b)
22. Balliet, D., Li, N.P., Macfarlan, S.J., Van Vugt, M.: Sex differences in cooperation: a meta-analytic review of social dilemmas. Psychol. Bull. **137**(6), 881–909 (2011)
23. Kopelman, S., Weber, J.M., Messick, D.M.: Factors influencing cooperation in commons dilemmas: a review of experimental psychological research. In: Ostrom, E.E., Dietz, T.E., Dolsak, N.E., Stern, P.C., Stonich, S.E., Weber, E.U. (eds.) The Drama of the Commons, pp. 113–156. National Academy Press, Washington, DC (2002)
24. Peshkovskaya, A., Babkina, T., Myagkov, M.: Social context reveals gender differences in cooperative behavior. J. Bioecon. **20**(2), 213–225 (2018). https://doi.org/10.1007/s10818-018-9271-5
25. Cubitt, R.P., Gachter, S., Quercia, S.: Conditional cooperation and betrayal aversion. J. Econ. Behav. Organ. (2015). https://doi.org/10.1016/j.jebo.2017.06.013
26. Frey, B.S., Meier, S.: Social comparisons and pro-social behavior: testing "conditional cooperation" in a field experiment. Am. Econ. Rev. **94**(5), 1717–1722 (2004)
27. Heldt, T.: Conditional cooperation in the field: cross-country skiers' behavior in Sweden. http://www.diva-portal.org/smash/get/diva2:521585/FULLTEXT01.pdf. Accessed 20 June 2020
28. Fischbacher, U., Gachter, S., Fehr, E.: Are people conditionally cooperative? Evidence from a public goods experiment. Econ. Lett. (2001). https://doi.org/10.1016/S0165-1765(01)00394-9
29. Lukinova, E., Babkina, T., Sedush, A., Menshikov, I., Menshikova, O., Myagkov, M.: Sociality is not lost with monetary transactions within social groups. CEUR-Workshop **1968**, 18–30 (2017)
30. Keser, C., Van Winden, F.: Conditional cooperation and voluntary contributions to public goods. Scand. J. Econ. **102**(1), 23–39 (2000)
31. Fischbacher, U.: z-Tree: Zurich toolbox for ready-made economic experiments. Exp. Econ. (2007). https://doi.org/10.1007/s10683-006-9159-4
32. Menshikov, I., Shklover, A., Babkina, T., Myagkov, M.: From rationality to cooperativeness: the totally mixed Nash equilibrium in Markov strategies in the iterated Prisoner's dilemma. PLoS ONE **12**(11), e0180754 (2017)

33. Kline, R., Seltzer, N., Lukinova, E., Bynum, A.: Causal responsibility, asymmetric opportunity and inequality in anthropogenic climate change (2014). http://sites.duke.edu/2014bmp/files/2014/10/Kline_Seltzer_Lukinova_Bynum.pdf.   Accessed 20 June 2020

34. Chartishvili, A.G., Kozitsin, I.V., Goiko, V.L., Saifulin, E.R.: On an approach to measure the level of polarization of individuals' opinions, pp. 1–5. IEEE (2019)

35. Kozitsin, I.V., et al.: Modeling political preferences of Russian users exemplified by the social network Vkontakte. Math. Models Comput. Simul. **12**(2), 185–194 (2020). https://doi.org/10.1134/S2070048220020088

36. Kozitsin, I.V., Marchenko, A.M., Goiko, V.L., Palkin, R.V.: Symmetric convex mechanism of opinion formation predicts directions of users' opinions trajectories, pp. 1–5. IEEE (2019)

Finite Games with Perturbed Payoffs

Vladimir A. Emelichev[1] and Yury V. Nikulin[2(✉)]

[1] Faculty of Mechanics and Mathematics, Belarusian State University,
220030 Minsk, Belarus
vemelichev@gmail.com
[2] Department of Mathematics and Statistics, University of Turku,
20014 Turku, Finland
yurnik@utu.fi

Abstract. The parametric concept of equilibrium in a finite cooperative game of several players in a normal form is introduced. This concept
is defined by the partitioning of the players into coalitions. In this situation, two extreme cases of this partitioning correspond to the Pareto
optimal outcome and the Nash equilibrium outcome, respectively. The
parameter space of admissible perturbations in such problem is formed
by a set of additive matrices, with two arbitrary Hölder norms specified
independently in the outcome and criterion spaces. The analysis of quasistability for a generalized optimal outcome under the perturbations of
the linear payoff function coefficients is performed. The limiting level of
such perturbations is found.

Keywords: Post-optimal analysis · Multiple criteria · Quasistability
radius · Parametric optimality

1 Introduction

The rapid development of various branches of information technology in economy and various social spheres, an important feature of which is their integrity,
high complexity and the presence of undefined factors, requires the creation of
adequate developments in the relevant areas of system analysis, management and
operations research. One of the main problems arising in this direction remains
the problem of making reasonable multi-purpose decisions in the conditions of
conflict. One of the effective tools of modeling such processes is the apparatus
of the mathematical game theory.

The goal of the game-theoretic model is to find classes of outcomes that
are rationally coordinated in terms of possible actions and interests of participants (players) or a group of participants (coalitions). For each game in normal
form, coalitional and non-coalitional equilibrium concepts (principles of optimality) are used, which usually lead to different game outcomes. In the theory of
non-antagonistic games there is no single approach to the development of such
concepts. The most famous one is the concept of the Nash equilibrium [1,2], as

© Springer Nature Switzerland AG 2020
N. Olenev et al. (Eds.): OPTIMA 2020, CCIS 1340, pp. 158–169, 2020.
https://doi.org/10.1007/978-3-030-65739-0_12

well as its various generalizations related to the problems of group choice, which is understood as the reduction of various individual preferences into a single collective preference.

In this paper, we introduce a parametrization of the equilibrium concept of a finite game in normal form. The parameter of this parameterizations is the method of dividing players into coalitions, in which the two extreme cases (a single coalition of players and a set of single-player coalitions) correspond to the Pareto optimal outcome and the Nash equilibrium outcome. Here, we study the type of stability of the game under consideration to perturbations of the parameters of the player payoff functions, which is a discrete analog of the Hausdorff upper semicontinuity property [3] of a multi-valued mapping that maps any set of game parameters to the corresponding set of all generalized equilibrium outcomes. As a result of the parametric analysis, the formula for the radius of quasistability of the coalition game was found under the assumption that arbitrary norms are specified in the two-dimensional space of game parameters.

The paper is organized as follows. In Sect. 2, we formulate parametric optimality and introduce basic concepts along with the notation. Section 3 contains some auxiliary statements about norms and four lemmas used later for the proof of the main result. In Sect. 4, we formulate and prove the main result regarding the quasistability radius. Section 5 lists most important corollaries. Section 6 contains numerical examples about bi-matrix games illustrating the main results.

2 Main Definitions and Notation

Consider a game of several players in normal form, where every player $i \in N_n = \{1, 2, \ldots, n\}$, $n \geq 2$ is choosing an action (strategy) $x_i \in \mathbf{R}$ to play from the finite set X_i, $2 \leq |X_i| < \infty$. The outcome of the game is a realization of the strategies chosen by all the players. Given a set of all possible outcomes of the game

$$X = \prod_{j \in N_n} X_j \subset \mathbf{R}^n,$$

for each player $i \in N_n$ we define a linear payoff function

$$f_i(x) = C_i x, \ i \in N_n,$$

where C_i is the i-th row of a square matrix $C = [c_{ij}] \in \mathbf{R}^{n \times n}$, $x = (x_1, x_2, \ldots, x_n)^T$, $x_j \in X_j$, $j \in N_n$. We assume all players try to maximize own payoffs simultaneously:

$$Cx = (C_1 x, C_2 x, \ldots, C_n x)^T \to \max_{x \in X}. \tag{1}$$

Since individual objectives are usually conflicting, a certain parameterized optimality principle will be introduced later.

A non-empty subset $J \subseteq N_n$ is called a coalition of players. For a coalition J and game outcome $x^0 = (x_1^0, x_2^0, \ldots, x_n^0)$ we introduce a set

$$V(x^0, J) = \prod_{j \in N_n} V_j(x^0, J)$$

where

$$V_j(x^0, J) = \begin{cases} X_j & \text{if } j \in J, \\ \{x_j^0\} & \text{if } j \in N_n \backslash J. \end{cases}$$

Thus, $V_j(x^0, J)$ is the set of outcomes that are reachable by coalition J from the outcome x^0. It is clear that $V(x^0, N_n) = X$ and $V(x^0, k) = X_k$ for any x^0.

In the space $\mathbf{R}^k$ of arbitrary dimension $k \in \mathbf{N}$ we introduce a binary relation that generates the Pareto optimality principle.

$$y \prec_P y' \Leftrightarrow y \leq y' \ \& \ y \neq y',$$

where $y = (y_1, y_2, ..., y_k)^T \in \mathbf{R}^k$, $y' = (y_1', y_2', ..., y_k')^T \in \mathbf{R}^k$. The symbol $\overline{\prec}$, as usual, denotes the negation of the relation $\prec$.

Let $s \in N_n$, and let $N_n = \bigcup\limits_{r \in N_s} J_r$ be a partition of the set N_n into s nonempty sets (coalitions), i.e. $J_r \neq \emptyset$, $r \in N_s$, and $p \neq q \Rightarrow J_p \cap J_q = \emptyset$. For this partition, we introduce a set of $(J_1, J_2, ..., J_s)$-efficient outcomes according to the formula:

$$G^n(C, J_1, J_2, \ldots, J_s) = \{x \in X : \\ \forall r \in N_s \ \forall x' \in V(x, J_r) \ (C_{J_r} x \, \overline{\prec}_P \, C_{J_r} x')\}, \tag{2}$$

where C_{J_r} is a submatrix of matrix C consisting of rows that correspond to players in coalition J_r. Sometimes for brevity, we denote this set by $G^n(C)$.

Thus, preference relations between players within the same coalition is based on Pareto dominance. Obviously, any N_n-efficient outcome $x \in G^n(C, N_n)$ ($s = 1$, i.e. all players are united in one coalition) is Pareto optimal, i.e. efficient outcome to game (1). Therefore, the set $G^n(C, N_n)$ is the Pareto set:

$$P^n(C) = \{x \in X : \forall x' \in X \ (Cx \, \overline{\prec}_P \, Cx')\}.$$

In the other extreme case, when $s = n$, $G^n(C, \{1\}, \{2\}, ..., \{n\})$ is a set of the Nash equilibria [1,2]. This set is denoted by $NE^n(C)$. Thereby, we have

$$NE^n(C) = \{x \in X : \not\exists k \in N_n \ \not\exists x' \in X \ (C_k x < C_k x' \ \& \ x_{N_n \backslash \{k\}} = x'_{N_n \backslash \{k\}})\}.$$

We assume that the game is such that it has at least one Nash equilibrium.

It is easy to see that rationality of the Nash equilibrium is that no player can individually deviate from the own equilibrium strategy choice while others keep playing their equilibrium strategies. Strict axioms regarding perfect and common (shared) knowledge are assumed to be fulfilled [4].

Thus, we have just introduced a parametrization of the equilibrium concept for a finite game in normal form. The parameter s of this parameterizations is the partitioning of all the players into coalitions $J = (J_1, J_2, ..., J_s)$, in which the two extreme cases (a single coalition of players and a set of single-player coalitions) correspond to finding the Pareto optimal outcomes $P^n(C)$ and the Nash equilibrium outcomes $NE^n(C)$, respectively.

Denoted by $Z^n(C, J_1, J_2, \ldots, J_s)$, the game consists in finding the set $G^n(C, J_1, J_2, \ldots, J_s)$. Sometimes for brevity, we use the notation $Z^n(C)$ for this problem.

Without loss of generality, we assume that the elements of partitioning $N_n = \bigcup_{r \in N_s} J_r$ be defined as follows:

$$J_1 = \{1, 2, \ldots, t_1\},$$
$$J_2 = \{t_1 + 1, t_1 + 2, \ldots, t_2\},$$
$$\ldots$$
$$J_s = \{t_{s-1} + 1, t_{s-1} + 2, \ldots, n\}.$$

For any $r \in N_s$, let C^r denote a square submatrix of size $|J_r| \times |J_r|$, consisting of those matrix C elements locates at the crossings of rows and columns with numbers J_r, and let $P^{|J_r|}(C^r)$ is the Pareto set of the problem

$$C^r z \to \max_{z \in X_{J_r}},$$

where $z = (z_1, z_2, \ldots, z_{|J_r|})^T$, and X_{J_r} is a projection of X onto J_r, i.e.

$$X_{J_r} = \prod_{j \in J_r} X_j \subset \mathbf{R}^{|J_r|}.$$

In particular case $s = 1$, we have $P^n(C) = G^n(C, N_n)$. It is evident that all matrices $C^{(r)}$, $r \in N_s$, form a diagonal block matrix C.

Due to the fact that the payoff linear functions $C_i x$, $i \in N_n$ are separable, the following equality is valid:

$$G^n(C, J_1, J_2, \ldots, J_s) = \prod_{r=1}^{s} P^{|J_r|}(C^r). \tag{3}$$

Perturbation of the elements of the matrix C is imposed by adding matrices B taken from $\mathbf{R}^{n \times n}$. Thus, the perturbed problem $Z^n(C + B)$ has the form

$$(C + B)x \to \max_{x \in X},$$

and the set of its $(J_1, J_2, \ldots, J_s)$-efficient outcomes is $G^n(C + B, J_1, J_2, \ldots, J_s)$.

In the space of game outcomes $\mathbf{R}^k$, $k \geq 2$, we define an arbitrary Hölder's norm l_p, $p \in [1, \infty]$, i.e. by the norm of the vector $a = (a_1, a_2, \ldots, a_k)^T \in \mathbf{R}^k$ we mean the number

$$\|a\|_p = \begin{cases} \left(\sum_{j \in N_k} |a_j|^p \right)^{1/p} & \text{if } 1 \leq p < \infty, \\ \max\{|a_j| : j \in N_k\} & \text{if } p = \infty. \end{cases}$$

The norm of the matrix $C \in \mathbf{R}^{k \times k}$ with the rows C_i, $i \in N_k$, is defined as the norm of a vector whose components are the norms of the rows of the matrix C. By that, we have

$$\|C\|_{pq} = \left\| \left(\|C_1\|_p, \|C_2\|_p, \ldots, \|C_k\|_p \right) \right\|_q,$$

where l_q, $q \in [1, \infty]$, is another Hölder's norm, i.e. l_p may differ from l_q in general case.

For an arbitrary number $\varepsilon > 0$, we define the set of perturbing matrices

$$\Omega(\varepsilon) = \left\{ B \in \mathbf{R}^{n \times n} : \|B\|_{pq} < \varepsilon \right\}.$$

Following [5], the *quasistability radius* of the game $Z^n(C, J_1, J_2, \ldots, J_s)$, $n \geq 2$, (called T_4-stability radius in the terminology of [6,7]) is the number

$$\rho = \rho_{pq}^n(J_1, J_2, \ldots, J_s) = \begin{cases} \sup \Xi & \text{if } \Xi \neq \emptyset, \\ 0 & \text{if } \Xi = \emptyset, \end{cases}$$

where

$$\Xi = \left\{ \varepsilon > 0 : \forall B \in \Omega(\varepsilon) \quad \left(G^n(C) \subseteq G^n(C + B) \right) \right\}.$$

Thus, the quasistability radius of the game $Z^n(C)$ determines the limit level of perturbations of the elements of the matrix C that preserve optimality of all the outcomes of the set $G^n(C)$ of the original problem $Z^n(C)$ but new extreme outcomes are allowed to arise in the perturbed problem $Z^n(C + B)$. The game $Z^n(C)$ is called *quasistable* if the quasistability radius is positive.

3   Auxiliary Statements and Lemmas

In the outcome space $\mathbf{R}^n$ along with the norm l_p, $p \in [1, \infty]$, we will use the conjugate norm l_{p^*}, where the numbers p and p^* are connected, as usual, by the equality

$$\frac{1}{p} + \frac{1}{p^*} = 1,$$

assuming $p^* = 1$ if $p = \infty$, and $p^* = \infty$ if $p = 1$. Therefore, we further suppose that the range of variation of the numbers p and p^* is the closed interval $[1, \infty]$, and the numbers themselves are connected by the above conditions.

Further we use the well-known Hölder's inequality

$$|a^T b| \leq \|a\|_p \|b\|_{p^*} \tag{4}$$

that is true for any two vectors $a = (a_1, a_2, \ldots, a_n)^T \in \mathbf{R}^n$ and $b = (b_1, b_2, \ldots, b_n)^T \in \mathbf{R}^n$.

Lemma 1. *For any $p \in [1, \infty]$ the following formula holds*

$$\forall b \in \mathbf{R}^n \ \forall \sigma > 0 \ \exists a \in \mathbf{R}^n$$

$$\left(|a^T b| = \sigma \|b\|_{p^*} \ \& \ \|a\|_p = \sigma \right).$$

Proof. It is well-known (see e.g. [8]) that Hölder's inequality becomes an equality for $1 < p < \infty$ if and only if

a) one of a or b is the zero vector;
b) the two vectors obtained from non-zero vectors a and b by raising their components' absolute values to the powers of p and p^*, respectively, are linearly dependent (proportional), and sign $(a_i b_i)$ is independent of i.

When $p = 1$, inequality (4) transforms into the following inequality:

$$\left| \sum_{i \in N_n} a_i b_i \right| \le \max_{i \in N_n} |b_i| \sum_{i \in N_n} |a_i|.$$

The last inequality holds as equality if, for example, b is the zero vector or if $a_j \ne 0$ for some j such that $|b_j| = \|b\|_\infty \ne 0$, and $a_i = 0$ for all $i \in N_n \backslash \{j\}$.

When $p = \infty$, inequality (4) transforms into the following inequality:

$$\left| \sum_{i \in N_n} a_i b_i \right| \le \max_{i \in N_n} |a_i| \sum_{i \in N_n} |b_i|.$$

The last holds as equality if, for example, b is the zero vector or if $a_i = \sigma \, \mathrm{sign} \, (b_i)$ for all $i \in N_n$ and $\sigma > 0$. $\square$

Directly from (3), the following lemma follows.

Lemma 2. *The outcome $x \in X$ is $(J_1, J_2, \ldots, J_s)$-efficient, i.e.*

$$x \in G^n(C, J_1, J_2, \ldots, J_s)$$

if and only if for any index $r \in N_s$

$$x_{J_r} \in P^{|J_r|}(C^r).$$

Hereinafter, x_{J_r} is a projection of vector $x = (x_1, x_2, \ldots, x_n)^T$ on coordinate axes of $\mathbf{R}^n$ with coalition numbers J_r.

The norm $\| \cdot \|$ defined in space $\mathbf{R}^n$ is called monotone if for any vectors $y, y' \in \mathbf{R}^n_+$ inequality $y \le y'$ implies $\|y\| \le \|y'\|$. It is well-known (see e.g. [8]) that all Hölder's norms l_p, $p \in [1, \infty]$ are monotone.

Hereinafter, a^+ is a projection of a vector $a = (a_1, a_2, \ldots, a_k) \in \mathbf{R}^k$ on a positive orthant, i.e.

$$a^+ = [a]^+ = (a_1^+, a_2^+, \ldots, a_k^+),$$

where $+$ implies positive cut of vector a, i.e.

$$a_i^+ = [a_i]^+ = \max\{0, a_i\}.$$

Lemma 3. *Given $x \notin G^n(C + B, J_1.J_2, \ldots, J_s)$, $B \in \Omega(\varphi)$, and $\varphi > 0$, then there exist $k \in N_s$ and $z^0 \in X_{J_k}$ such that inequality*

$$\|[C^k(x_{J_k} - z^0)]^+\|_q < \varphi \, \|x_{J_k} - z^0\|_{p^*} \tag{5}$$

holds.

Proof. Since $x \notin G^n(C + B, J_1.J_2, \ldots, J_s)$, due to lemma 2, there exists index $k \in N_s$ such that

$$x_{J_k} \notin P^{|J_k|}(C^k + B^k).$$

Thus, due to the fact of external stability of the Pareto set (see e.g. [9]), there exists vector $x^0 \in P^{|J_k|}(C^k + B^k)$ such that

$$(C^k + B^k)x_{J_k} \leq (C^k + B^k)z^0.$$

Then we have

$$(C_i^k + B_i^k)(x_{J_k} - z^0) \leq 0, \ i \in J_k.$$

So, due to inequalities (4), we obtain

$$[C_i^k(x_{J_k} - z^0)]^+ \leq \|B_i^k\|_p \|x_{J_k} - z^0\|_{p*}, \ i \in J_k. \tag{6}$$

Let $J_k = \{i_1, i_2, \ldots, i_v\}$, $1 \leq i_1 \leq i_2 \leq \cdots \leq i_v \leq n$. Taking into consideration (6) as well as the property of l_q-norm monotonicity, we deduce inequalities (5).

$$\|[C^k(x_{J_k} - z^0)]^+\|_q = \|[C_{i_1}^k(x_{J_k} - z^0)]^+, [C_{i_2}^k(x_{J_k} - z^0)]^+, \ldots, [C_{i_v}^k(x_{J_k} - z^0)]^+\|_q \leq$$

$$\|B^k\|_{pq} \|x_{J_k} - z^0\|_{p*} \leq \|B\|_{pq} \|x_{J_k} - z^0\|_{p*} < \varphi \|x_{J_k} - z^0\|_{p*}. \qquad \square$$

Lemma 4. *Assume $\emptyset \neq J_k \subseteq N_n$, $k \in N_s$, $z^0, z \in X_{J_k}$, $z^0 \neq z$. Let matrix C^k with rows C_i^k, $i \in J_k$, and vector η with positive elements η_i, $i \in J_k$, be such that inequality*

$$[C_i^k(z^0 - z)]^+ < \eta_i \|z^0 - z\|_{p*}, \ i \in J_k \tag{7}$$

holds, Then for any $\varepsilon > \|\eta\|_q$ there exists matrix

$$B^k \in \mathbf{R}^{|J_k| \times |J_k|}$$

such that

$$z^0 \notin P^{|J_k|}(C^k + B^k),$$

$$\|B_i^k\|_p = \eta_i, \ i \in J_k,$$

$$\|B^k\|_{pq} < \varepsilon.$$

Proof. Let $\varepsilon > \|\eta\|_q$. According to Hölder's inequality (4), for any matrix $D^k \in \mathbf{R}^{|J_k| \times |J_k|}$ with rows D_i^k, $i \in J_k$, the following inequalities are valid:

$$D_i^k(z^0 - z) \leq \|D_i^k\|_p \|z^0 - z\|_{p*}, \ i \in J_k.$$

Therefore, for any index $i \in J_k$ due to lemma 1 there exists matrix B^k with rows B_i^k, $i \in J_k$ such that

$$B_i^k(z^0 - z) = -\eta_i \|z^0 - z\|_{p*},$$

$$\|B_i^k\|_p = \eta_i, \ i \in J_k.$$

From the above expressions taking into account (7), we deduce

$$(C_i^k + B_i^k)(z^0 - z) \leq [C_i^k(z^0 - z)]^+ - \eta_i \|z^0 - z\|_{p*} < 0, \ i \in J_k,$$

i.e. $z^0 \notin P^{|J_k|}(C^k + B^k)$, where $\|B^k\|_{pq} = \|\eta\|_p < \varepsilon$. $\qquad \square$

4 Quasistability Radius

For the game $Z^n(C, J_1, J_2, \ldots, J_s)$, $n \geq 2$, $s \in N_n$ and any $p, q \in [1, \infty]$, we define

$$\varphi = \varphi_{pq}^n(J_1, J_2, \ldots, J_s) = \min_{x \in G^n(C)} \min_{r \in N_s} \min_{z \in X_{J_r} \setminus \{x_{J_r}\}} \frac{\|[C^r(x_{J_r} - z)]^+\|_q}{\|x_{J_r} - z\|_{p^*}}.$$

It is obvious that $\varphi \geq 0$.

Here we formulate the main result of this work. The analytical formula specified in the main theorem below provides and enumerative way of calculating the quasistability radius, i.e. extreme level of perturbations preserving all outcomes of the original (non-perturbed) game $Z^n(C, J_1, J_2, \ldots, J_s)$,. Why is it important to have information about quasistability radius? First, if the radius of quasistability is not equal to zero, it determines the equilibriums not only to the original game, but also to a series of games with parameters located in the vicinity of the radius. Second, for a number of particular cases one can potentially build an algorithm for finding radii that uses and continues the same procedures that were involved in the game solving, which actually means that the radius could be potentially calculated along with the equilibrium of the game.

Theorem 1. *For any $p, q \in [1, \infty]$, $C \in \mathbf{R}^{n \times n}$, $n \geq 2$ and $s \in N_n$, the quasistability radius of the game $Z^n(C, J_1, J_2, \ldots, J_s)$ is expressed by the formula:*

$$\rho = \rho_{pq}^n(J_1, J_2, \ldots, J_s) = \varphi_{pq}^n(J_1, J_2, \ldots, J_s).$$

Proof. First, we prove the inequality $\rho \geq \varphi$. For $\varphi = 0$, this inequality is obvious. Let $\varphi > 0$. Then according to the definition of the number φ, we have

$$\forall x \in G^n(C) \quad \forall r \in N_s \quad \forall z \in X_{J_r} \setminus \{x_{J_r}\}$$

$$\left(\|[C^r(x_{J_r} - z)]^+\|_q \geq \varphi \|x_{J_r} - z\|_{p^*} > 0 \right). \tag{8}$$

Assume the opposite, i.e. assume that $\rho < \varphi$. Hence, there exists matrix $B \in \Omega(\varphi)$ such that $x \neq G^n(C + B, J_1, J_2, \ldots, J_s)$. Thus, due to lemma 3 there exist index $k \in N_s$ and vector $z^0 \in X_{J_k}$ such that strict inequality (5) holds. Then it contradicts to inequality (8), i.e. we proved that $\rho \geq \varphi$.

Further, we prove that $\rho \leq \varphi$. Let $\varepsilon > \varphi$ and $\Theta > 1$ be such that

$$\varepsilon > \Theta \varphi > \varphi.$$

Then according to the definition of the number φ, we have

$$\exists x^0 \in G^n(C) \quad \exists k \in N_s \quad \exists z \in X_{J_k} \setminus \{x_{J_k}^0\}$$

$$\left(\|[C^k(x_{J_k}^0 - z)]^+\|_q = \varphi \|x_{J_k}^0 - z\|_{p^*} \right). \tag{9}$$

Then there exists vector η with positive components $\eta_i \in J_k$ such that

$$[C_i^k(x_{J_k}^0 - z)]^+ < \Theta[C_i^k(x_{J_k}^0 - z)]^+ = \eta_i \|x_{J_k}^0 - z\|_{p^*}, \ i \in J_k.$$

$$\|\eta\|_q = \Theta\varphi < \varepsilon.$$

Using lemma 4, we deduce that there exists matrix B^k of size $|J_k| \times |J_k|$ with rows B_i^k, $i \in J_k$ such that

$$x_{J_k}^0 \notin P^{|J_k|}(C^k + B^k),$$

$$\|B_i^k\|_p = \eta_i, \ i \in J_k,$$

$$\|B^k\|_{pq} = \|\eta\|_q = \Theta\varphi < \varepsilon.$$

Summarizing all the above and taking into account lemma 2, we conclude that for $x^0 \in G^n(C)$ the following statement is true

$$\forall \varepsilon > \varphi \ \exists B \in \Omega(\varepsilon) \ (x^0 \notin G^n(C + B, J_1, J_2, \ldots, J_s)),$$

where $B \in \mathbf{R}^{n \times n}$ is a matrix composed of the elements of matrix B^k located at the crossings of rows J_k and columns J_k, and zeroes otherwise. Hence, for any $\varepsilon > \varphi$ it holds that $\rho < \varphi$. So, we have $\rho \leq \varphi$. □

5   Corollaries

Corollary 1. *For any* $p, q \in [1, \infty]$, $C \in \mathbf{R}^{n \times n}$, $n \geq 2$ *the quasistability radius of the game* $Z^n(C, N_n)$ *consisting in finding the Pareto set* $P^n(C)$ *is expressed by the formula:*

$$\rho_{pq}^n(N_n) = \min_{x \in P^n(C)} \ \min_{x' \in X \setminus \{x\}} \frac{\|[C(x - x')]^+\|_q}{\|x - x'\|_{p^*}}.$$

Corollary 1 implies that the game $Z^n(C, N_n)$ is quasistable if and only if the Pareto set $P^n(C)$ coincides with the Smale set [10] $S^n(C)$ defined as:

$$S^n(C) = \{x \in P^n(C) : \ S^n(x, C) = \emptyset\},$$

where

$$S^n(x, C) = \{x' \in X \setminus \{x\} : \ Cx = Cx'\}.$$

From theorem 1 it follows

$$\rho_{pq}^n(\{1\}, \{2\}, \ldots, \{n\}) = \min_{x \in NE^n(C)} \ \min_{i \in N_n} \ \min_{z \in X_i \setminus \{x_i\}} \frac{\|[c_{ii}(x_i - z)]^+\|_q}{\|x_i - z\|_{p^*}}.$$

From here for any $x \in NE^n(C)$ and $z \in X_i \setminus \{x_i\}$ the equalities hold

$$\frac{\|[c_{ii}(x_i - z)]^+\|_q}{\|x_i - z\|_{p^*}} = \frac{\|c_{ii}(x_i - z)\|_q}{\|x_i - z\|_{p^*}} = |c_{ii}|.$$

So, we get the following result.

Corollary 2. *For any $p, q \in [1, \infty]$, $C \in \mathbf{R}^{n \times n}$, $n \geq 2$ the quasistability radius of the game $Z^n(C, \{1\}, \{2\}, \ldots, \{n\})$ consisting in finding the Nash set $NE^n(C)$ is expressed by the formula:*

$$\rho_{pq}^n(\{1\}, \{2\}, \ldots, \{n\}) = \min\{|c_{ii} : \ i \in N_n|\}.$$

Corollary 2 implies that the game $Z^n(C, \{1\}, \{2\}, \ldots, \{n\})$ is quasistable if and only if all the main diagonal elements of matrix C are different from zero. Theorem 1 also implies the following result.

Corollary 3. *The outcome $x^0 = (x_1^0, x_2^0, \ldots, x_n^0)^T$ of the game with matrix $C \in \mathbf{R}^{n \times n}$, $n \geq 2$ is the Nash equilibrium, i.e. $x^0 \in NE^n(C)$ if and only if the equilibrium strategy for each player $i \in N_n$ is as follows:*

$$x_i^0 = \begin{cases} \max\{x_i : \ x_i \in X_i\} \ if \ c_{ii} > 0, \\ \min\{x_i : \ x_i \in X_i\} \ if \ c_{ii} < 0, \\ x_i \in X_i \qquad\qquad if \ c_{ii} = 0. \end{cases}$$

Corollary 3 implies that the game $Z^n(C, \{1\}, \{2\}, \ldots, \{n\})$ is quasistable if and only if

$$|NE^n(C)| = 1.$$

6   Numerical Examples

Consider several examples of bi-matrix games with two players. Let $C \in \mathbf{R}^{2 \times 2}$ is a matrix with rows C_1 and C_2, and let $X_i \in \{0, 1\}$, $i \in N_2$, $x^{(1)} = (0, 0)^T$, $x^{(2)} = (0, 1)^T$, $x^{(3)} = (1, 0)^T$, $x^{(4)} = (1, 1)^T$. These examples illustrate different interrelations between quasistability radii for Nash and Pareto optimality principles. Set $p = q = \infty$. The payoff functions are written as

$$\begin{bmatrix} (C_1 x^{(1)}, C_2 x^{(1)}) \ (C_1 x^{(2)}, C_2 x^{(2)}) \\ (C_1 x^{(3)}, C_2 x^{(3)}) \ (C_1 x^{(4)}, C_2 x^{(4)}) \end{bmatrix}$$

Additionally, set (see corollaries 1 and 2)

$$\rho^2(P^2(C)) = \rho_{\infty\infty}^2(N_2) = \min_{x \in P^2(C)} \ \min_{x' \in X \setminus \{x\}} \ \max_{i \in N_2} \frac{\|C_i(x - x')\|_q}{\|x - x'\|_1}. \tag{10}$$

$$\rho^2(NE^2(C)) = \rho_{\infty\infty}^2(\{1\}, \{2\}) = \min\{|c_{ii}| : \ i \in N_2\}. \tag{11}$$

Example 1. Let

$$C = \begin{pmatrix} 2 & -6 \\ -2 & 1 \end{pmatrix}.$$

Then we have bi-matrix game $Z^2(C)$ with payoffs

$$\begin{bmatrix} (0, 0) & (-6, 1) \\ (2, -2) & (-4, -1) \end{bmatrix}.$$

Therefore,
$$P^2(C) = \{x^{(1)}, x^{(2)}, x^{(3)}\},$$
$$NE^2(C) = \{x^{(4)}\}.$$

It is evident that Pareto optimal outcome $x^{(1)}$ is not the Nash equilibrium. This type of game is known as prisoner's dilemma see (e.g. [4]). According to formulae (10) and (11), we have

$$\rho^2(P^2(C)) = \rho^2(NE^2(C)) = 1.$$

Example 2. Let
$$C = \begin{pmatrix} 2 & -1 \\ -1 & 0 \end{pmatrix}.$$

Then we have bi-matrix game $Z^2(C)$ with payoffs
$$\begin{bmatrix} (0,0) & (-1,0) \\ (2,-1) & (1,-1) \end{bmatrix}.$$

Therefore,
$$P^2(C) = \{x^{(1)}, x^{(3)}\},$$
$$NE^2(C) = \{x^{(3)}, x^{(4)}\}.$$

According to formulae (10) and (11), we have

$$\rho^2(P^2(C)) = \frac{1}{2}, \quad \rho^2(NE^2(C)) = 0.$$

Example 3. Let
$$C = \begin{pmatrix} 2 & 3 \\ 5 & 1 \end{pmatrix}.$$

Then we have bi-matrix game $Z^2(C)$ with payoffs
$$\begin{bmatrix} (0,0) & (3,1) \\ (2,5) & (5,6) \end{bmatrix}.$$

Therefore,
$$P^2(C) = NE^2(C) = \{x^{(4)}\}.$$

According to formulae (10) and (11), we have

$$\rho^2(P^2(C)) = 3, \quad \rho^2(NE^2(C)) = 1.$$

Example 4. Let
$$C = \begin{pmatrix} -2 & -1 \\ 1 & -3 \end{pmatrix}.$$

Then we have bi-matrix game $Z^2(C)$ with payoffs
$$\begin{bmatrix} (0,0) & (-1,-3) \\ (-2,1) & (-3,-2) \end{bmatrix}.$$

Therefore,
$$P^2(C) = \{x^{(1)}, x^{(3)}\},$$
$$NE^2(C) = \{x^{(1)}\}.$$

According to formulae (10) and (11), we have

$$\rho^2(P^2(C)) = 1, \quad \rho^2(NE^2(C)) = 2.$$

7   Conclusion

As a result of parametric analysis performed, the formula for the quasistability radius was obtained in a finite cooperative game of several players in a normal form with parametric optimality ranging from Pareto solutions to Nash equilibria in the case where criterion and solution spaces are endowed with various Hölder's norms.

One of the biggest challenges in this field is to construct efficient algorithms to calculate the analytical expression. To the best of our knowledge, there are not so many results known in that area, and moreover some of those results which have been already known, put more questions than answers. As it was pointed out in [11], calculating exact values of stability radii is an extremely difficult task in general, so one could concentrate either on finding easy computable classes of problems or developing general metaheuristic approaches.

References

1. Nash, J.: Equilibrium points in n-person games. Proc. Natl. Acad. Sci. **36**(1), 48–49 (1950)
2. Nash, J.: Non-cooperative games. Ann. Math. **54**(2), 286–295 (1951)
3. Aubin, J.-P., Frankowska, H.: Set-Valued Analysis. Birkhuser, Basel (1990)
4. Osborne, M., Rubinstein, A.: A Course in Game Theory. MIT Press, Cambridge (1994)
5. Emelichev, V., Nikulin, Yu.: On a quasistability radius for multicriteria integer linear programming problem of finding extremum solutions. Cybern. Syst. Anal. **55**(6), 949–957 (2019)
6. Sergienko, I., Shilo, V.: Discrete Optimization Problems. Problems, Methods Research. Naukova dumka, Kiev (2003)
7. Emelichev, V., Kotov, V., Kuzmin, K., Lebedeva, N., Semenova, N.: Stability and effective algorithms for solving multiobjective discrete optimization problems with incomplete information. J. Autom. Inf. Sci. **46**(2), 27–41 (2014)
8. Hardy, G., Littlewood, J., Polya, G.: Inequalities. University Press, Cambridge (1988)
9. Noghin, V.: Reduction of the Pareto Set: An Axiomatic Approach. Springer, Cham (2018). https://doi.org/10.1007/978-3-319-67873-3
10. Smale, S.: Global analysis and economics V: pareto theory with constraints. J. Math. Econ. **1**(3), 213–221 (1974)
11. Nikulin, Y., Karelkina, O., Mäkelä, M.: On accuracy, robustness and tolerances in vector Boolean optimization. Eur. J. Oper. Res. **224**, 449–457 (2013)

Optimization in Big Data Analysis Based on Kolmogorov-Shannon Coding Methods

Georgy K. Kamenev[1] ID, Ivan G. Kamenev[1,2](✉) ID, and Daria A. Andrianova[3]

[1] Federal Research Center of Computer Science and Control Dorodnitsyn Computing
Centre of the Russian Academy of Sciences, Moscow, Russia
igekam@gmail.com
[2] FGAEI HE National Research University "Higher School of Economics",
Moscow, Russia
[3] Yandex N.V., Moscow, Russia

Abstract. The article describes the optimization problem solving for
multidimensional and bulks Big Data (data with more than 10 characteristics and 10^8 observations or higher), as well as machine-generated
data of unlimited volume. It is difficult to analyze and visualize data
of such volume and complexity using traditional methods. In contrast
(and in addition) to machine learning methods widely used in Big Data
analysis, it is proposed to use stochastic methods of data sets' coding
and approximation using Kolmogorov-Shannon metric nets, which are
optimal for the entropy of the code. While adapting these methods, new
methods are proposed for metrics construction for characteristics with
nominal and ordinal scales.

Keywords: Multidimensional statistical analysis · General
population · Sociological sample · Metric net · Data visualization ·
Data analysis · Data optimization · Method of metric data analysis ·
Decisive minorities

1 Introduction

Optimization problems can be set either on some mathematical model that con-
nects a space of variables with systems of equations and computational algorithms or on some multidimensional data set. In such a set, each data unit
(observation) is characterized by several variables. The optimization problem, in
this case, is to find the optimal value of the functional given on the variables from
the set of observations, as well as a particular solution on which the functional
takes the optimal value. Theoretically, such problems can be solved by exhaustive
search (brute-forced) being discrete and finite, but in the case of Big Data, there
are several reasons why it is necessary to obtain and investigate the structure
(topological and metric) of the entire set of optimal and suboptimal solutions

The reported study was co-funded by RFBR, project number: 18-01-00465 a: Development of social data multi-dimensional metric analysis methods.

(observations). This might be necessary due to the format and quality of the Big Data used, as well as due to the goals of optimization research. In the first case, for example, for the optimal particular solution (observation), there may be no information on certain important attributes that the optimized functional does not depend on, thus such information must be obtained from other optimal or suboptimal observations. In the second case, for example, the researcher may be interested in the most complete analysis of the set of all optimal or suboptimal observations as a significant minority.

Using an applied financial and economic case, we will describe and demonstrate our proposed approach to solving optimization problems on multidimensional sets with a high level of complexity, including those that arise in the study of Big Data. We consider the task of finding a complete optimal or suboptimal solution on Big Data as on a fractal-like set in a multidimensional space of variables extended (if needed) by a functional, using the Kolmogorov-Shannon [3,4] ε-nets for approximation. In this case, the ε value characterizes the accuracy of optimization problem solving and should be chosen rather small. The complexity or entropy of the approximation problem in small is determined by the fractal (metric) dimension. The fractal dimension of Big Data in the extended space approximation with the required precision can be significantly smaller than the dimension of the space of Big Data variables. Therefore, to solve the optimization problem on the data fractal-like set, we use our own [5] stochastic methods for the construction of C. Shannon's (ε, δ)-nets whose convergence rate is determined by the fractal dimension of the approximated set.

Our methods were previously used separately (see, for example, [10,11]) and as united complex (the method of metric data analysis, MMDA) [12] for analyzing and visualizing model-generated data, i.e. searching for optimal solutions to a "black box" mapping, for example, a system of equations without its analytical solution. In contrast, this publication describes the adaptation of these methods to the problems of Big Data research, in which the model as a system of equations cannot be constructed, at least at the current stage of research.

Our coding and approximation methods allow us to use a direct (ε, δ)-coverage as a mathematical-statistical model, to study the properties of the model, and to formulate assumptions about the nature of dependencies in the data. Such research is necessary before the construction of the system of equations describing Big Data (the so-called "machine learning").

In one of the previous works, we tested MMDA on the example of a public Big Data array [15] used in various data analysis competitions. The next stage of the method's approbation is presented in this publication. We apply it to real practical problems (which also made us adapt and clarify some of the method's components). We use data from a large Russian payment system (for more information, see the section "Generation of Big Data array of payment transactions"). In this publication, we present the results of the MMDA application to this data flow related to real optimization problems solved by the company (mainly, the maximization of the average life-long profit from the client). Other results that are of interest to data science, but not directly related to optimization, are highlighted in separate publications.

2 Big Data as a Specific Data Type

There are several approaches to the definition of Big Data. Some sources define them only by the criterion of the amount of the information (number of records/observations) [6]. Others – by their structure's volume and complexity (data that does not fit into the logic of standard databases) [7]. And still, others refer to the technical/business origin [8] or large data arrays, that analysts are not able to process in traditional ways. We use a hybrid definition, meaning that Big Data is a data of technical origin (i.e. the data that occurs while the information is being recorded by the technical system), which cannot be processed by traditional methods due to their multidimensionality (a large number of variables/ characteristics), a large number of observations, and the erroneous records appearing among the observations.

Big Data is a sequence of records (rows, vectors), each of which contains a row's ID and information, represented by the number of characteristics. Big Data can be presented as a flow (so-called "log") or a data array. In fact, any array is a log's sample, for example, a log for the specific moment of time. In this paper it is information about the transactions of the payment system's clients.

Research on Big Data can be described by the following logical scheme:

1. Formation of the Big Data array (sampling it from the Big Data flow)
2. Automated preprocessing of the Big Data array
3. Automated statistical analysis of Big Data array
4. Automated modeling of Big Data flow

1. Since the Big Data array is by definition a selection from the General population (the data stream of a socio-technical system), the content of the Big Data array is initially determined by the selection method. It could be selection by records' numbers (including by date). In this case, the system's operation period is selected that, for the data logging structure, it is realistic to collect every record into array in the time available to the researcher. If the period is well-founded and sufficiently long, then such a study can be considered a census.

   Alternatively (as in this study), a randomized sample (random record numbers) may be used. The usage of randomized sampling is associated with some risks since many data streams have a non-uniform intensity (which leads to distortions in frequency analysis). However, randomized sampling allows us to cover longer periods and check the universality of flow features (their presence in different circumstances, including different times). Sampled Big Data array requires a statistical assessment of the quality of the sample (for metric analysis, completeness and reliability of the coverage that the sample forms).

2. Preprocessing involves converting the Big Data array to a technical format, interpreting and correcting gaps and errors, since the unformatted and distorted data are unavoidable in any real data flow. Preprocessing is often performed several times if new problems are identified in subsequent analysis stages.
3. The higher is the Big Data dimension, the less standard statistical methods are suitable for research: less informative and, most importantly, more resource-intensive. In particular, if the computational complexity of queries for a single characteristic is proportional to N, where N is the number of records, then the complexity of clustering analysis is proportional to at least N^2. A similar problem occurs in regression analysis concerning the number of characteristics. As a result, exploratory research on Big Data is traditionally limited to the simplest statistical functions, rather than studying their internal metric features (impossible combinations of characteristics, outliers, etc.).
4. The ultimate goal of Big Data research usually is to construct a system of equations and inequalities that describes (approximates) the entire data array. It should reflect the essential features of the data for the researcher. The types of equations are set by the researcher. A complex computer algorithms identify the model by selecting the coefficients of equations.

Unfortunately, the study of Big Data using algorithmic methods requires a clear statement of the problem and an intuitive understanding of the features of the data array, which are achieved by trial and error (often in the form of "championships" for specific problem's solving). Thus machine learning is widely used for solving well-formalized technical problems with limited multidimensionality (2–3 dimensions). Social data, as a rule, are more multidimensional, and algorithms calibration on low-cardinality samples is incorrect due to the heterogeneous intensity of the initial data flows.

The method we developed allows us to create a data model in the form of a metric net, rather than a system of equations. This makes it applicable to social Big Data. This model allows us to form a general understanding of the data features and correct any errors. This creates the necessary prerequisites for the construction of the model in the form of the forecasting equation system.

3   Method of Metric Data Analysis

The metric analysis is a methodological approach that includes the research of the data's topological and metric properties (the distribution of observations in the metric space of characteristics). This approach is mostly developed in the (Engineering) Sciences, where it is used to solve standardized problems in small-dimensional space (for example, in image recognition). Multidimensional metrics analysis, on the contrary, remains the area where the application of existing theoretical concepts is limitly developed, especially in the field of mathematical methods of Social Sciences.

In this study, we demonstrate the multidimensional metric analysis' capability for the optimization problem-solving in the Social Sciences, based on the method of Metric data analysis (MMDA) [9], simultaneously adapting it for Big Data. The method is based on the approach to MCDM developed in [1,2] and based on approximation and visual analysis of the sets. The approximation is based on constructing of ε-nets [4] and (ε, δ)-nets [3]. This complex method allows us to identify sets of optimal and suboptimal solutions for various optimization problems, to localize them metrically, and to research the properties of a localized subset.

Let us first describe the MMDA approach for the study of implicitly defined sets of the "black box" type. Let the set X of possible observation belongs to the decision space Rn. The objective functions given by the non-linear mapping $f \colon \mathrm{R}^n \to \mathrm{R}^m$ that relates decisions to m criterion. Therefore, the set $f(X)$ is usually non-convex. The set $X \subset \mathrm{R}^n$ is assumed to be compact. Then, the set $f(X)$ is compact, too. The method described herein is based on the selection of random observations, i.e. uniformly distributed random points x from the set X. These outputs filtered by the approximation algorithm provide the basis for evaluation and display of the set $Z = f(X)$.

Geometrically, this means that a set is approximated by a collection of simple figures (such as balls or cubes of appropriate dimension), whose diameters are taken smaller to achieve a higher approximation accuracy. A metric $\varepsilon - net$ of a set is its subset such that any point of a set lies at a distance of at most ε from the subset. If "balls" are constructed in the considered spatial metric around the points of a metric ε-net, then they cover the entire approximated set forming ε-covering. This corresponds to the set approximated by a collection of balls in the Euclidean metric; or by a collection of cubes in the Chebyshev metric (as in this research); or any other relevant metrics. For each point of such a collection, we can quickly find a "true" point of the set (the nearest point of the metric ε-net) separated from it by a distance of at most ε in the considered metric. Moreover, for smaller ε, the true points are closer to the points of the approximated set and there are fewer "redundant" points in the approximating collection of balls, but a greater number of points have to be constructed in the net. The optimal complexity of ε-nets constructing is characterized using the metric ε-entropy [4].

The collection of centers of the balls (boxes) is named the approximation base and is denoted by T of M elements. The set $(T)_\varepsilon$, that is, the collection of ε-neighborhoods of the points in T, forms then the *approximation* of the set Z. To find criterion points that can provide a good approximation of Z, a global sampling of the set X is carried out. To be more precise, uniformly distributed random points x of the set X are generated. Then, their outputs $f(x)$ are computed, and a small number of them (see below) are selected as the centers of boxes that form the approximation. The length of the edges of the boxes must be specified in such a way that the resulting system of boxes approximates the set Z with a desired degree of accuracy. Collections of two- or free-criterion slices of the approximation can be then displayed reasonably fast.

Let T and ε be given. To evaluate the quality of an approximation of Z by $(T)_\varepsilon$, two indicators are used that describe the deviation of $(T)_\varepsilon$ from Z and the deviation of Z from $(T)_\varepsilon$. For the first indicator, the value of ε is used. The base T belongs to Z, and the deviations of the points of $(T)_\varepsilon$ from Z are not greater than ε. The *completeness* of the approximation is characterized by the fact that only a small part (the part with a small measure) of the approximated set lies outside the constructed collection of balls with centers at points of the net. To estimate the completeness η of the ε-covering (the measure of the covered set) we use a test sample consisting of N points from the pre-image set X and check the sampling fraction η^* of its outputs that belong to the tested covering $(T)_\varepsilon$. The collection of balls with accuracy ε and completeness η forms so-called (ε, δ)-coverings, $\delta = 1 - \eta$ [3].

In our approach the reliability of (ε, δ)-covering (the probability of proximity of η and η^*) depends on the value of N. More precisely, let us denote probability defined on the space of samples of the volume N as P. Then for measurable mapping the reliability $P\{\eta > \eta^* - \delta\}$ of $\eta^* - \delta$ to be lower estimate of the completeness η of a given coverage is evaluated by the following result [5]:

Theorem 1. $P\{\eta > \eta^* - \triangle\} \geq \chi(\triangle, N), \chi(\triangle, N) = 1 - \exp(2N\triangle^2)$

Due to this theorem it follows that in case $\eta^* = 1$ it holds

$$P\{\delta < \triangle\} \geq \chi(\triangle, N) \tag{1}$$

This result characterizes the sample size N needed to construct an approximation with a given completeness η and reliability χ: $\chi = 0.95$, $N = 150$, $\delta = 0.1$ $\chi = 0.95$, $N = 600$ $\delta = 0.05$ $\chi = 0.999$, $N = 30000$, $\delta = 0.01$ $\chi = 0.99$, $N = 100000$, $\delta = 0.005$.

Bounded implicitly defined sets (images of maps) can be approximated by applying adaptive stochastic techniques based on the Deep Holes method (DHM) [5]. Specifically, for a set defined implicitly by a nonlinear "black-box" mapping, these techniques are used to construct a covering that approximates it with prescribed accuracy ε and completeness η.

Stochastic DHM algorithm. It is assumed that the value of N has already been specified (or controlled in process of approximation). In the framework of this algorithm, some stopping rules may be applied. For example, the rule based on the maximal number of points in the approximating base $M = M_{\max}$ and the rule testing the condition $\eta^* = 1$. It was proven that for any compact set X and any measurable (for example continuous) mapping f, with a properly specified number N this algorithm can construct an approximation base T that satisfies (1) in a finite number of iterations for any values of ε, χ and δ).

Proof. Iteration of the Algorithm. The current approximation base T is assumed to be constructed on the previous iterations of the algorithm.

1. Generate the test sample: N random uniformly distributed points from the set X and compute their outputs.
2. Compute the fraction η^* of the test sample with outputs in $(T)_\varepsilon$.
3. Stop if $\eta^* = 1$ (then $P\{\triangle > \delta\} \geq \chi(\triangle, N)$) or if the number of points M in T is equal $M_{\max}$ (then $P\{1 - \eta^* + \triangle > \delta\} \geq \chi(\triangle, N)$). Otherwise, augment the current approximation base T by the most distant from T output among the N sample points and go to step 1.

Imagine that we have constructed an approximation $(T)_\varepsilon$ of the set Z. The approximation has a simple explicit description as a system of boxes with centers that belong to the set Z and edges of length 2ε. Since the number of boxes is relatively small, 2D or 3D slices of the approximation can rapidly be computed and displayed by computer graphics. For each point $(T)_\varepsilon$ it is possible to find the nearest point of the approximation base T and to reconstruct the corresponding decision. In practice, the approach is at its best with up to five criteria. In the case of more than three criteria, 2D and 3D visualization are to be used with scroll-bars that help to fix or to control the values (or ranges) of the other criteria and to study the influence of them on the three-criterion pictures (Interactive Decision Maps (IDM) technique [2]). Note that though the approximation process may take a fairly long time, the visualization procedure can be carried out by the decision-maker in real-time without waiting after the request.

Now we will present an adaptation of the MOD technique for the task of studying Big Data extraction from a certain General population: a finite or infinite set G of elements (real, hypothetical, potential) of the metric space R^d, containing big data $X, X \subseteq G \subset R^d$.

Two data analysis problem formulations characterizing the relationship between the available data set X and the general population G: (1) X coincides with $G, X \equiv G$; (2) X is a proper subset of $G, X \subset G$.

In formulation (2) the following assumptions are postulated:

- The general population G is a bounded subset of R^d.

This assumption is always satisfied with real physical and social phenomena.

- X is an independent sample from G of the volume M obtained on some probability measure μ_G.

Failure to comply with this assumption or the lack of data makes it impossible to make conclusions about the general population G from the X set analysis. In this case, the problem transforms into a formulation (1).

BD implementation of MMDA consist in:

- construction of $T - \varepsilon-$ or (ε, δ)-net and approximation $S = (T)_\varepsilon - \varepsilon-$ or (ε, δ)-covering of X using DHM (for (ε, δ)-net with probability measure μ_X);
- (in formulation (2)): testing of the completeness δ^G of S approximation on General population G or on its control sample;
- to make visual, topological and/or metric analysis of S structure;
- in optimization case – to solve (multy)optimization problem on S.

Let $M(e, A)$ be the maximal cardinality (the number of elements) of ε-disjoint set (ε-packing) for A and dmA - fractal dimension of A. According to [5].

Theorem 2. *Complexity of the DHM approximation for BD problems is* $O(M(\varepsilon/2, X))M$ *for* (ε, δ)-*covering*, dmX $<<$ d, *asymptotically* $O(\varepsilon^{-dmX})M)$;

Theorem 3. *Complexity of the DHM approximation is* $O(M(\varepsilon/2, X)N$ *for* ε-*covering*, $dmX << d, N << M$, *asymptotically* $O(\varepsilon^{-dmX}))N$

The full algorithm of the Metric data analysis method includes ([13]):

1. Collection of data about characteristics: for Big Data, this is a sample survey of general populations of data that are side-products of digital technology systems.
2. Selection of characteristics that are relevant for the study.
3. Creation of a multidimensional space (metrics).
4. General population' reconstruction by (ε, δ)-covering approximation; determining the reliability, accuracy, and completeness of the representation of the General population by the constructed approximation using its test sample.
5. Structural and visual analysis of (ε, δ)-covering approximation: identification of types, classes, and separate groups based on metrical proximity.
6. Generalization of characteristics and study of the behavior of selected classes: while using this method for optimization, the goal is to identify a class of optimal and/or suboptimal solutions as decisive minorities.

4   Generation of Big Data Array of Payment Transactions

In this research, we analyze the Big Data flow of a payment system, which is a collection of records about characteristics of payments. The flow capacity V is about 10^8 order of magnitude annually, and each payment in the system has more than 20 key characteristics. This makes this Big Data flow unavailable for analysis by traditional methods, except the simplest statistical calculations (minimum, maximum, average, correlation of two characteristics).

There are 15 relevant characteristics in the array among the others:

1. *Transaction ID* (unique serial number);
2. *ID of the payer* who made this transaction (let's denote the number of payers in the flow as V);
3. *unixtime* of transaction (seamless continuous time scale);
4. *type* of payer (individual, organization, etc.);
5. *purpose* of the transaction (i.e. receiver type);
6. *method* of the execution of the transaction (i.e. ;
7. *interface* where the command was given to execute the transaction;
8. *value* of the transaction;
9. *profit* from the transaction;
10. *bonuses* accrued by the transaction;

11. *total number* of transactions of the payer who made this transaction;
12. *total value* of transactions of the payer who made this transaction;
13. *total profit* from of transactions of the payer who made this transaction (let's denote it as U_i, $i \in 1 \ldots V$);
14. time from the *registration* of the payer;
15. time from the *first payment* by the payer.

There is an optimization task for this set: we are interested in factors that maximize the total profit from the given client for the entire time of his observation in the payment system. What is the optimal combination of payment characteristics by which we can assume that the payer will bring the company the greatest profit? Note that the transactional costs and the user's maintaining cost are negligible compared to the constant costs of the payment infrastructure's maintaining. The profitability of the client is determined mainly by whether he chooses or not to perform transactions for which a significant commission is set.

Since it is not possible to convert all the logs from different servers into a single array, we have to use sample survey methods. Since we are not interested in statistical properties of individual characteristics (average/minimum/maximum, which are elementary calculated in SQL), but in combinations of characteristics, standard sociological samples are not enough – they are unreliable. Based on the available computing power, we decided to create a sample based on the payment id with a $4 * 10^6$ order of magnitude. Ids of the transactions for the sample were determined by the python module *random*, using *random.seed*() [16].

The calculation of the metric dimension of the data [14] allows us to evaluate the quality of this sample. It was shown that such a sample size is not sufficient to make reasonable judgments about the combinations of all characteristics at once, but it is sufficient to operate with 13 characteristics that are essential for this study: their metric dimension tends to 1.03.

We consider a problem in which the Dig Data array X does not coincide with the General population (data stream) G, and G is limited metrically. Then it is necessary to test the reliability of conclusions about the properties of the General population G from the properties of the data set X. To do this, we build (ε, δ)- nets of $n(\varepsilon)$ elements and calculate their metric (fractal) dimension:

$$dm \leq d, where : n(\varepsilon) \sim (1/\varepsilon^{dm}), \varepsilon \to 0 \qquad (2)$$

The metric dimension test [14] has shown that this sample size is not sufficient to make informed judgments ($\varepsilon \sim 0.2$) about the combinations of all characteristics (as dimensions of a multidimensional space), but it is convenient to operate with $\approx$5 characteristics with $\varepsilon \sim 0.0001$. (Note that each combination of characteristics had to be tested independently).

5   Preprocessing of Big Data Array of Payment Transactions

We performed some preprocessing procedures, the details of which are omitted because they relate to data analysis and are not of interest for optimization methodology: 1. All the records that meet one of the following conditions (errors, failures, and unique operations) were removed to a separate array:

- the quantitative characteristic of the transaction is a spike (metrically isolated from other observations);
- the qualitative characteristic of a transaction occurs less frequently than 2–3 times in the sample;
- the transaction is single, i.e. the payment system was not able to match it with the payer.

2. Since the user id is obviously confidential (it allows you to calculate a specific contractor), so it was used for the logs integration, but it was not included in the array.
3. All quantitative variables were normalized to the maximum value in the array to protect confidentiality and trade secrets (thus, they take values from 0 to 1 and, in some cases, negative values).
4. For variables with nominal scales, ranks that do not occur in the array were checked by logs and excluded. Note also that the presence of variables with nominal scales imposes additional restrictions on the used (ε, δ)-nets: ε must be at least $1/(n-1)$ to avoid distortion of the metric, where n is the number of ranks of the nominal scale.

We also skip the details of research on combining overlapping classifications, since they are not directly related to the optimization problem being solved.

The number of observations in the array after preprocessing was 3210599. Note that the reduction of the array's size is not the goal of preprocessing. The sample size from flow G has already been determined so that our algorithms can be implemented to it in an acceptable time. Preprocessing allows excluding anomalous observations from the array that distort the metric. A meaningful study of metric features and identification of compact minorities is possible during exploratory metric analysis performed using MMDA (approximation by (ε, δ)-coverage).

6   Exploratory Metric Analysis

Exploratory approximations aim to identify significant patterns in the data in a limited time. Full visualization of a two-dimensional or three-dimensional projection of an array of 3–4 million records can take 5–10 min (this time is spent every time the projection axes and/or metric constraints are changed). As a result, such an array is suitable for visualizing already known patterns, but not for detecting them (which requires testing a set of hypotheses). We built exploratory approximating metric nets for several key sets of characteristics to solve this problem. If the visualization is completely unusable (for example, if the sample exceeds 10 million observations), then it is advisable to construct a high-quality approximating metric net ($\varepsilon \approx 0$).

As a result, it was found that the array's metric is significantly distorted by several key clients: organizations that perform an order of magnitude more outgoing transactions than regular users (primarily financial organizations and service aggregators). Such corporate clients are served at individual rates under special agreements, so it is not appropriate to take into account their observations while studying the behavior of ordinary clients. Despite their small number, they create the majority of the sample due to a large number of transactions. After their elimination, maximums of normalized characteristics changed: for example, for the total number of transactions, it became 0.1. The majority of the persisting payments are characterized by the total value of payments below 0.01 and the total profit below 0.02.

The exploratory analysis showed a significant metric relationship between the target (optimized) characteristic (total profit) and such characteristics as profit (from a given transaction), transaction value, purpose, and accrued bonuses. On the contrary, the most obvious hypothesis that the total profit from the client is proportional to the duration of his work in the system was not confirmed in general (see Fig. 1).

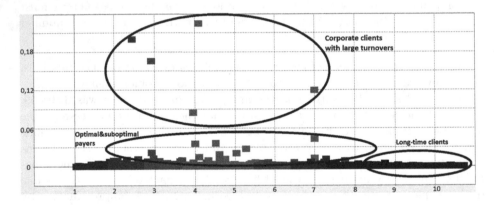

Fig. 1. Payments whose payers are characterized by a high *total profit* (ordinate), depending on the time from the payer's *registration* (abscissa) and from his *first payment* (color) (original sample's visualization). (Color figure online)

7 Revelation of Optimal and Suboptimal Solutions

Let's explore three-dimensional projections that visualize metric features of a set of optimal and suboptimal observations. Optimizational task is to maximize average (life-long) *total profit* from payers:

$$F \to \max, F = \sum_{i=1}^{V} U_i/V \tag{3}$$

To achieve this task, assuming payers' independence, the payment system ought to increase the P_i, which requires an understanding of the properties of payers with maximum or near-maximum total profit.

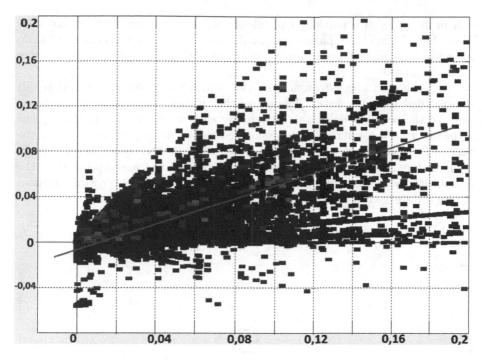

Fig. 2. Payments whose payers are characterized by a high *totalprofit* (color, from above, from 0.2 to 0). depending on the *value* of a single payment (abscissa) and the *profit* from a single payment (ordinate) (original sample's visualization). (Color figure online)

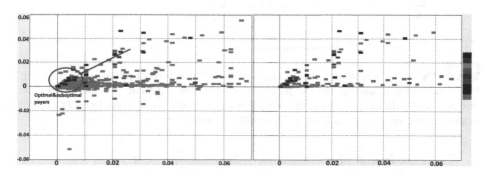

Fig. 3. A (on the left). Same projection zoomed in (payment *value* and payment *profit* less than 0.008) (high quality approximating metric net). B (on the right). Same projection excluding payments whose payers brought zero or negative $total_profit$

It can be seen that observations characterized by a high total profit are localized in compact minorities by the parameters of the value and profit from a particular payment (see Fig. 2, 3). They are found on several lines (corresponding to the interest rates of the payment commission), which are characterized by the high profitability of a particular payment and a proportionally large payment value. At the same time, payers who made the payment for the largest value do not make a significant overall profit, especially if the payment itself also does not make a profit.

We recall that single-transactions (for which the payer is not registered) are excluded from the array. Therefore, this picture shows that a significant profit to the payment system is brought by customers considering the type of payments with a significant commission fee as the main purpose of using the system. Profitable customers still make payments of other types, but the main payment type dominates. It is possible to evaluate the assumption that the payers with the highest total profit tend to make payments for a certain purpose and by certain methods, in projection on these variables (see Fig. 4).

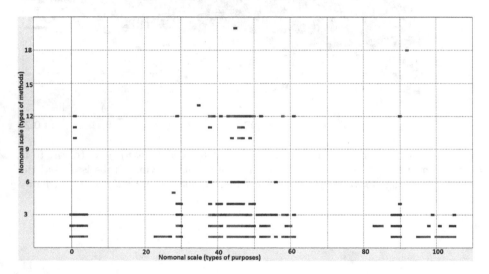

Fig. 4. Payments whose payers are characterized by a medium *total profit* (color, below 0.2), depending on the *purpose* and *method* of payment (nominal scales on abscissa and ordinate accordingly), excluding payments whose payers brought zero or negative total profit (original sample's visualization). (Color figure online)

Significant overall profit is generated by customers who make transactions for *purposes*: large online stores, services owned by the payment system's owner, various online games, and the stock market. The payment methods for high-profit clients are: from an account with some type of card linked to account. Some high-profitable clients are business users who use some specific methods for transitional payments for third payment systems (similar to the large corporate clients excluded from the analysis, but on a smaller scale). Another minority of high-profitable clients makes mostly transfers to the other accounts.

Users with the lowest profitability have transactions to the same online stores as profitable clients, but paying by cash. This may indicate a leak in profitability accounting. Also, all users using mobile commerce (payment from the phone

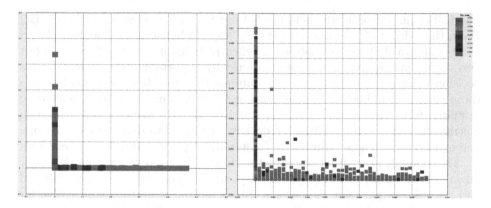

Fig. 5. A (on the left). The amount of *bonuses* accrued for the payment (abscissa) and the $total_{p}rofit$ from the payer who made the payment (ordinate), with *value* of a payment (color). B (on the right). The same projection excluding payments whose payers brought zero or negative $total_{p}rofit$, and payers who did not receive *bonuses*, zoomed in (on a scale up to 0.01). (Color figure online)

balance) turned out to be low-profitable. Among the low-profitable clients, there is a group of people using mainly nonprofit-by-Law transactions (transport fees, communal services, taxes, etc.). Low profitability is mostly located by the payment method, rather than its purpose. The users of third-party cards and mobile payment systems in various offline stores are usually moderately profitable.

The impact of the bonus program on the total profit from the client appears to be highly non-trivial. The purpose of the bonus program is to make one-time customers permanent and encourage them to use payment services more actively, including services with a high commission. However, at first glance, the total profit from the client and the value of bonuses accrued for the operation are anticorrelated (see Fig. 5A). High overall profits are earned from customers who do not earn bonuses, and high bonuses are awarded to customers who do not make a profit. This topological structure calls into question the effectiveness of the bonus program. Note, however, that the area of the potential effectiveness of the bonus program still exists (lower-left corner of the picture). Let's zoom it in (see Fig. 5B). It can be seen that bonuses affect the behavior of "normal" users with the turnover typical for individuals, although not every one of them and on a limited scale. Some (but not all) users are still encouraged to make profit-making transactions by accruing bonuses.

8   Conclusions and Significance

The presented research describes the logic of optimization problem solving by metric analysis. In this example, we researched the problem of single-criteria optimization based on the Big Data array in the field of Social Sciences and using the method of Metric data analysis (MMDA). Given the limited applicability of classical optimization methods in the field of Social Sciences, especially for Big Data, the subject of this study is not only applied but also methodological.

It is shown that metric analysis can effectively identify not only optimal but also suboptimal solutions, which is of practically important in the Social Sciences decision-making (allowing to find not only for the "first-best" solution but also those close to it). The methodological improvements in the MMDA (including sampling from the log data, building search, and high-precision approximating metric nets) allow us to research the data by criteria with multiple local optima, and by factors with non-natural scale (nominal, i.e. positive integer of categories).

The example shows that the target variable (the total profit from the payer during his life in the payment system) depends primarily on what type (method and purpose) of payment the payer is interested in, being the main motive for the payment system's usage. This allows the commercial service to make more informed decisions about the system's monetization and stimulation programs' focuses. The metric analysis shows that the set of (sub)-optimal solutions, i.e. payers who brought the greatest profit to the company, has the following characteristics: these are people who make often (but not very often) not too large payments, for which high commissions are set. It is also shown that payers tend to choose one scenario for the payment system's usage, i.e. make payments of a certain type with a certain commission. This is confirmed by the fact that lines corresponding to the payment system commission rates occur not only on the *"value − profit"* projection (where their presence is trivial), but also on the *total value − total profit* projection. Accordingly, the payment system must make efforts to promote payments with a high commission rate among payers who are only familiar with commission-free payments (for example, by issuing options for the full commission discounts for the first payments of the certain types). Such solutions will help payers to try out new types of payments that are beneficial to the payment system, meaning their transition to the (previously specified) (sub)optimal type.

The presented data also illustrates the typical social science case where Big Data can be correctly interpreted only after the separation of crucial minorities. For example, a minority of large corporate clients significantly distort payment statistics due to a large number of transactions performed, while a minority of payments made by individuals react fundamentally differently to the bonus program. The system can only be set up to maximize the total profit from the client after these two groups of payers are separated based on the metric features of their data (with a side result in clarifying the classification of corporate clients for the security service).

References

1. Lotov, A.V., Bushenkov, V.A., Kamenev, G.K.: Feasible Goals Method: Mathematical Foundations and Environmental Applications. The Edwin Mellen Press, Lewiston (1999)

2. Lotov, A.V., Bushenkov, V.A., Kamenev, G.K.: Interactive Decision Maps. Approximation and Visualization of Pareto Frontier. Applied Optimization, vol. 89. Kluwer Academic Publishers, Boston/Dordrecht/New York/London (2004)
3. Shannon, C.: The mathematical theory of communication. Bell Syst. Tech. J. **27–28**, 379–423, 623–656 (1948)
4. Kolmogorov, A.N., Tikhomirov, V.M.: Epsilon-entropy and Epsilon-capacity of sets in functional spaces. Adv. Math. Sci. **25**(2), 3–86 (1959)
5. Kamenev, G.K.: Approximation of completely bounded sets by the deep holes method. Comput. Math. Math. Phys. **41**(11), 1667–1675 (2001)
6. Leetaru, K. How Do We Define Big Data And Just What Counts As A Big Data Analysis? https://www.forbes.com/sites/kalevleetaru/2019/01/09/how-do-we-define-big-data-and-just-what-counts-as-a-big-data-analysis/#13d886081b66
7. Ward, J.S., Barker, A.: Undefined by data: a survey of big data definitions. arXiv preprint arXiv:1309.5821 (2013)
8. Davenport, T.H., Barth, P., Bean, R.: How big data is different. MIT Sloan Manag. Rev. **54**(1), 1–5 (2012)
9. Kamenev, G.K., Kamenev, I.G.: Multidimensional statistical sets and their metric analysis. Proc. Inst. Syst. Anal. Russ. Acad. Sci. CC FRC CSC RAS **68**(2), 30–33 (2018). https://doi.org/10.14357/20790279180207
10. Bolgov, M.V., Buber, A.L., Lotov, A.V.: Support for making strategic decisions on the water supply of the lower Volga River based on the Pareto frontier visualization. Sci. Tech. Inf. Process. **45**(5), 297–306 (2018). https://doi.org/10.3103/S0147688218050027
11. Kamenev, G.K., Sarancha, D.A., Polyanovsky, V.O.: Investigation of the class of one-dimensional unimodal mappings obtained in the modeling of the lemming population. Biophysics **63**(4), 596–610 (2018). https://doi.org/10.1134/S0006350918040097
12. Kamenev, G.K., Kamenev, I.G.: Discrete-dynamic modeling of governance for human capital. Matem. Mod. **32**(6), 81–96 (2020). https://doi.org/10.20948/mm-2020-06-06
13. Kamenev, G.K., Kamenev, I.G.: Metric analysis of multidimensional sociological samples [Metricheskij analiz mnogomernyh sociologicheskih vyborok] In: Pospelov, I.G., et al. (eds). Proceedings of the conference "Modeling of Coevolution of Nature and Society: Problems and Experience. To the 100-th Anniversary from the Birthday of Academician N.N. Moiseev", pp. 198–209. FRC CSC of RAS, Moscow (2017)
14. Kamenev, I.G., Andrianova, D.A.: Mathematical and statistical methods of pre-processing and exploratory analysis of Big social Data on the example of payments stream analysis. In: Proceedings of the 62nd all-Russian Scientific Conference of MIPT, Applied Mathematics and Computer Science, pp. 40–42. MIPT, Moscow (2019)
15. Andrianova, D.A., Kamenev, I.G.: The method of metric data analysis in big data in transport streams research. In: Proceedings of the IX Moscow International Conference of Operations Research, ORM 2018, vol. 2, pp. 506–510 (2018)
16. Python 3.8.3 documentation: random - generate pseudo-random numbers. https://docs.python.org/3/library/random.html
17. Berezkin, V.E., Kamenev, G.K., Lotov, A.V.: Program for visualization of multidimensional Pareto-frontier in non-convex multi-criteria optimization problems (PFV-II) (2019)

Construction of a Technologically Feasible Cutting with Pierce Points Placement Constraints

Tatiana Makarovskikh$^{(\boxtimes)}$ ⓘ and Anatoly Panyukov ⓘ

South Ural State University, Chelyabinsk, Russia
{Makarovskikh.T.A,paniukovav}@susu.ru
http://www.susu.ru

Abstract. The routing problem arising in cutting is considered. This task is to find the path of the cutting tool that satisfies the technological restrictions and the pierce points placement constraints. Since the time for piercing significantly affects the duration of the cutting process, it is necessary to reduce both the number of pierce points and the distance between successive fragments of the path. This research is devoted to routing problems in plane graphs, which are homeomorphic images of cutting plans. The route covering all the borders of the cut parts determines the path of the cutting tool. The technological constraints are: (1) the absence of intersection of the route internal faces of any initial part with the edges of its remaining part (OE-condition); (2) self-intersections of the cutting path prohibition (NOE-condition); (3) the initial vertices of the covering chains must allow piercing ($PPOE$-condition). The report presents the polynomial algorithm for constructing a cutting route that satisfies the introduced restrictions and consists of a minimum number of chains. The first two classes are considered and well studied earlier. In this paper we considered the class of $PPOE$-routes in plane graphs, i.e. routes with restrictions on choice of starting vertices (corresponding the pierce points on a sheet) of covering chains. We proved the necessary and sufficient conditions of these routes existence. Also we developed the polynomial time algorithm *PPOE-routing* for obtaining of such routes and proved its correctness.

Keywords: Optimal route · Cutting · Routing problem · Polynomial algorithm

1 Introduction

Cutting processes are used to shape engineering materials with complex shapes and strict design and performance functional requirements. The process is used

The work was supported by Act 211 Government of the Russian Federation, contract No. 02.A03.21.0011. The work was supported by the Ministry of Science and Higher Education of the Russian Federation (government order FENU-2020-0022).

ⓒ Springer Nature Switzerland AG 2020
N. Olenev et al. (Eds.): OPTIMA 2020, CCIS 1340, pp. 186–197, 2020.
https://doi.org/10.1007/978-3-030-65739-0_14

for cutting, drilling, marking, welding, sintering and heat treatment processes. Applications of sheet cutting include aerospace, automobile, shipbuilding, electronic and nuclear industries. The modern technologies of thermal cutting (gas, laser, plasma) as well as water jet cutting allow to implement cutting plans with combined cuts.

If we consider laser cutting, the intense laser light is capable to melt almost all materials. Laser cutting is a thermal energy based non-contact process, therefore does not require special fixtures and jigs to hold the workpiece. In addition, it does not need expensive or replaceable tools to produce mechanical force that can damage thin, intricate and delicate work pieces. The effectiveness of cutting depends on the thermal, optical and mechanical properties of materials [7].

Cutting is a well established process in manufacturing with high quality requirements. The cutting quality is influenced by many factors, for example, the laser power, focal position and gas pressure may be concerned for laser cutting. In order to achieve the desired cutting result in terms of quality and process velocity, it is generally required to identify suitable factor combinations for the specific laser cutting system in use and the workpiece at hand. Finding such a combination is often time consuming and cost intensive [1].

Such factor combinations may be formalised as different conditions and restrictions. To define a path of the cutter satisfying these conditions one needs solve some optimisation problems. For example, the paper [23] is devoted to the CPDP (Cutting Path Determination Problem), which consists in determining the optimal path for cutting according to a given cutting plan with one or more tools. The authors assume that there are two obvious restrictions: 1) all parts must be cut out; 2) none of the cut out parts should require further cuts, i.e. OE (Ordered Enclosing) constraint [14] is fulfilled. To solve the CPDP problem, more detailed statements are known: GTSP (General Traveling Salesman Problem [2–6,9,10,22]), CCP (Continuous Cutting Problem Point) [21], ECP (Endpoint Cutting Problem), see [11,17], and ICP (Intermittent Cutting Problem, see [14]). A new approach for minimizing both cutting path and heat accumulation in laser cutting process is presented in paper [9]. The proposed algorithm was based on a memetic algorithm combining a powerful genetic algorithm with an adaptive large neighbourhood search. The CPDP may be modeled and solved in accordance with generalized travelling salesman problem. Note that ECP and ICP allow the combination of the parts borders, which reduces material waste, cutting length and idle lengths (see [4]). The problems of reducing material waste and maximizing the combination of the contours fragments of the cut out parts are solved at the stage of the cutting plan design.

Despite the noted advantages of the computer technologies ECP and ICP most publications are currently devoted to the development of GTSP and CCP technologies, which use obvious cutting path algorithms consisting in contour-by-contour cutting.

The development of ECP and ICP computer technologies are considered, for example, in papers by [14,17,20]. The polynomial algorithms for OE routing (when the part cut off from a sheet does not require further cuts) are given there.

For industrial enterprises related by their activity to the tasks of cutting sheet material, there is a need to use CAD/CAM systems for the technological preparation of cutting processes. Taking into account the capabilities of modern equipment for cutting parts from sheet material allows you to make cutting plans that allow combining the contours of the cut parts, which reduces material waste, cutting length, and the number of idle passes. Algorithms for cutting plans design for tasks that allow combination of cuts do not fundamentally differ from algorithms that do not allow combination. However, the algorithms for finding the paths of the cutting tool movement are fundamentally different. Therefore, the development of algorithms for finding the route of the cutting tool for cutting plans that allow the combination of the contours of the cut parts is an open task.

Our research is devoted to routing problems in plane graphs, which are homeomorphic images of cutting plans. The route covering all the borders of the cut parts determines **the path of the cutting tool**. The technological constraint is the absence of intersection of the internal faces of any route initial part with the edges of its remaining part. When constructing manipulator control systems using an undirected graph as a model of cutting plan we may display various elements of the manipulator trajectory by it. In this case, problems of constructing routes that satisfy various constraints arise.

In this paper we consider the task to find the path of the cutting tool that satisfies the technological restrictions and constraints on the pierce points placement. Since the time for piercing significantly affects the duration of the cutting process, it is necessary to reduce both the number of pierce points and the distance between successive fragments of the path. We present the polynomial algorithm for constructing a cutting route that satisfies the introduced restrictions and consists of a minimum number of chains.

The first section of our paper is devoted to the statement of the problem and its formalization in terms of graphs and routes.

In the second section we discuss different necessary conditions of solution existence.

In the third section we present the polynomial time algorithm for constructing a route with fixed pierce points and discuss its correctness.

2   Statement of the Problem

Let us consider the problem that arises in the case of violation of restrictions on the location of pierce points. We consider the case of determining the cutting tool trajectory during cutting, when it is necessary to leave space for the implementation of the piercing. In addition, the piercing time significantly affects the duration of the cutting process. Therefore, we have the problem of determining the possibility of cutting for a given cutting plan, as well as the task of minimizing the number of pierce points.

In our past papers [14,15] we consider the plane S as a cutting sheet, the model of a cutting plan is a plane graph $G = (V, F, E)$ with outer face f_0 defined on S. The edges $e \in E$ of plane graph are the fragments of contours, which are the plane non-intersecting Jordan curves. If a curve is not closed then their bounding points are the vertices $v \in V$ of graph G. The closed Jordan curves correspond to loops in graph G. So, if we consider plane graph as homeomorphic image of the cutting plan, then the trajectory of a cutter be a route in graph. Then the number of pierce points may be defined as a number of edge-disjoint chains covering this graph. According to theorem proved in [12] the number of pierce points is not less than $|V_{odd}|/2$.

Let us consider cutting plans in Fig. 1. We admit that these cutting plans have combined cuts. This means that piercing is possible only for vertices incident to outer face (in common, for faces allowing piercing).

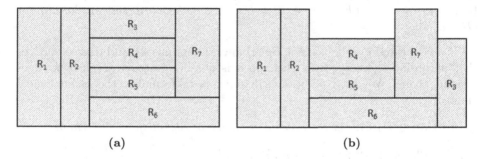

Fig. 1. The examples of realizable and non-realizable cutting plans for cutting using combined cuts technology

Homeomorphic images of these cutting plans are shown in the Fig. 2.

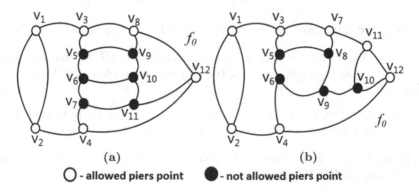

Fig. 2. The examples of graphs satisfying the realizable and non-realizable cutting plans (see Fig. 1) for cutting using combined cuts technology

So, the cutting plan in Fig. 1(b) is realizable at the point of view of considered cutting technology, and cutting plan in Fig. 1(a) is not. This cutting plan has lack of pierce points to cut the inner rectangles R_4 and R_5. The homeomorphic image 2(b) of the cutting plan in Fig. 1(b) allows placement of pierce point at outer contour, and the homeomorphic image 2(a) of the cutting plan in Fig. 1(a) does not allow placement of pierce point at inner regions.

This problem may be formalized as following [16].

Let faces $F_{in}(G) \subset F(G)$ allow piercing. Let odd degree vertices $V_{in}(G) \subset V(G)$ are incident to face $F_{in}(G)$. If obtained route is an OE-route and all starting points of covering chains belong to $V_{in}(G)$ then this route may be used as a base for constructing a route of cutter moving. Such routes are called $PPOE$-routes [13, 16].

Let us to remind the definition of OE-route.

Definition 1 ([19]). *Let chain* $C = v_1 e_1 v_2 e_2 \ldots v_k$, $0 < k \leq |E(G)|$ *so that* $\mathrm{Int}(v_1 e_1 v_2 e_2 \ldots e_l) \cap E(G) = \emptyset$, $1 \leq l \leq k$ *be called **ordered enclosing** (or OE) chain.*

If a connected graph G is not Eulerian and contains $2k$ odd degree vertices then according to Listing-Luke theorem it is possible to cover graph by k edge-disjoint chains. We need additional definitions to formalize the technological claims.

Definition 2 ([18]). *Let the ordered sequence of edge-disjoint OE-chains*

$$C^0 = v^0 e_1^0 v_1^0 e_2^0 \ldots e_{k_0}^0 v_{k_0}^0, \ C^1 = v^1 e_1^1 v_1^1 e_2^1 \ldots e_{k_1}^1 v_{k_1}^1, \ldots,$$
$$C^{n-1} = v^{n-1} e_1^{n-1} v_1^{n-1} e_2^{n-1} \ldots e_{k_{n-1}}^{n-1} v_{k_{n-1}}^{n-1}$$

covering graph G and such that

$$(\forall m : m < n), \ \left(\bigcup_{l=0}^{m-1} \mathrm{Int}(C^l) \right) \cap \left(\bigcup_{l=m}^{n-1} C^l \right) = \emptyset$$

*be called **cover with ordered enclosing** (or OE-cover for short).*

Routes realizing OE-cover represent the ordered set of OE-chains and contain additional idle passes (edges) between the end of current chain and beginning of the next [15].

Definition 3. *[16] Let a chain $C = v_1 e_1 v_2 e_2 \ldots v_k$ be called **PPOE-chain** if it is the OE-chain starting at vertex $v_1 \in V_{in}(G)$.*

Definition 4. *Let **PPOE-cover** of graph G be such an OE-cover of graph G consisting only of PPOE-chains.*

Definition 5. *Let **Eulerian PPOE-cover** be the minimal cardinality ordered sequence of edge-disjoint PPOE-chains in plane graph G.*

Graphs in Figs. 2(a) and (b) are the images of cutting plans in Figs. 1(a) and (b) correspondingly. Vertices V_{in} are presented by white circles. These vertices allow placement of pierce points near them. Vertices marked by black circles do not allow placement of pierce point near them. Thus, there exists Eulerian $PPOE$-cover for graph in Fig. 2(b). For example, it may be the following: $C_1 = v_1v_3v_5v_6v_9v_8v_5$, $C_2 = v_3v_7v_8$, $C_3 = v_7v_{11}v_{10}v_9$, $C_4 = v_{11}v_{12}v_{10}$, $C_5 = v_{12}v_4v_6$, $C_6 = v_4v_2v_1v_2$. Graph in Fig. 2(a) has no $PPOE$-cover.

Let us consider the necessary and sufficient conditions for $PPOE$-cover existence and the algorithm for obtaining this cover in polynomial time.

3   Necessary Conditions of $PPOE$-Route Existence

The problem of determining the feasibility of a cutting plan can be formulated as determining the existence of an Euler $PPOE$-cover for a plane graph that is a homeomorphic image of the corresponding cutting plan. In accordance with the existing restrictions, we can formulate the following necessary condition for the existence of a $PPOE$-covering [13].

Proposition 1. *[13] If G is a plane graph with $2k$ of odd vertices, and G has Eulerian $PPOE$-cover then $|V_{in}(G)| \geq k$.*

For example, graph in Fig. 3(a) can not be covered by $PPOE$-chains. This graph has 8 odd vertices, i.e. it may be covered minimum by four OE-chains, nevertheless only three of them may start at vertices available as starting points for $PPOE$-chains. As for graph in Fig. 3(b), it has four vertices that can be starting ones for $PPOE$-chains and the same number of end-points. But this graph also has no $PPOE$-cover.

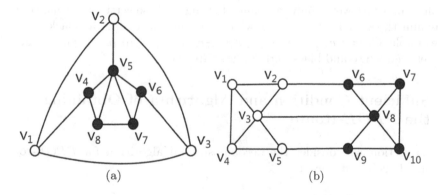

Fig. 3. The examples of graphs with no $PPOE$-cover

The routes realizing $PPOE$-cover may be presented as the ordered set of $PPOE$-chains connected by supplementary edges between the end of the current chain and beginning of another one. Such transitions allow to get a bipartite oriented graph $D = (V_{in} \cup V_{out}- > V_{in}, E)$ for which V_{in} be the set of odd vertices used to be the beginning of a chain (pierce points); V_{out} be the set of odd vertices used to be only the ends of obtained chains (end points).

Proposition 2. *[13] The PPOE-cover exists if for a mixed graph $G \cup D$ there exists cycle which edges $E(D)$ (if $e \in E(D)$ then $e \notin E(G)$)belong to*

$$\{(v, u): \ v \in V_{out} \cup V_{in}, \ u \in V_{in}\}.$$

Proof. As soon as $PPOE$-cover is a partial case of OE-cover then it has the directed cycle consisting of edges $e \in E(G)$ and edges of matching M on the set of vertices $V_{odd} \in G$ (vertices $V_{in} \cup V_{out}$). As soon as $PPOE$-cover consists of $PPOE$-chains then the edges of matching M should be passed in the direction $V_{out} \cup V_{in} - > V_{in}$. In this case these edges will correspond to arcs of D.

Thus, for $PPOE$-cover existence we need existence of such a cycle for a mixed graph $G \cup D$ where all supplementary edges $e \in E(D)$ be the arcs from $V_{out} \cup V_{in}$ in V_{in}.

Proposition 3. *[13] The PPOE-cover for a connected graph G exists is the cardinality of minimal $\{V_{in}, V_{out}\}$-cut is not greater than $|V_{out}|$.*

Proof. Let $PPOE$-cover exists for graph G. Let, nevertheless, cardinality of $\{V_{in}, V_{out}\}$-cut be less than $|V_{out}|$. As soon as no one of $PPOE$-chains of a cover cannot start at $u \in V_{out}$ then the cover consists of not more than $|V_{out}|$ routes from $v \in V_{in}$ to $u \in V_{out}$. Then some of these routes may turn up edge-disjoint. This contradicts with definition of $PPOE$-cover.

As an example we may consider graph in Fig. 3(b). $PPOE$-chain may start only from white vertex. This graph has ten odd vertices, hence, we need minimum five white vertices. The considered graph has such five vertices, nevertheless, this graph cannot be covered only by chains starting in these vertices. We may obtain minimum three chains starting at white vertices and ending at black vertices, for example, $C_1 = v_5v_9v_8v_{10}v_9$, $C_2 = v_3v_8v_7v_6v_8$, $C_3 = v_4v_3v_2v_6$. Thus, minimal cut between white and black vertices has three edges.

4   Sufficient Condition and Algorithm of Obtaining the *PPOE*-Route

In this section we consider the decompositional algorithm for $PPOE$-routes defining if such routes exists.

Algorithm *PPOE-routing*
Require: plane graph $G(V, E)$ defined by functions $v_k(e)$, $l_k(e)$, $r_k(e)$, $f_k(e)$, $k = 1, 2$ and rank(e) [15]; sets $V_{out}, V_{in} \subset V$.
Ensure: $PPOE$-cover of graph $G(V, E)$:

$$\tilde{C}_1, \tilde{C}_2, \ldots \tilde{C}_{|V_{out}|}, C_{|V_{out}|+1}, \ldots, C_M.$$

Step 1. Construct the network $N(V, A)$ (i.e. directed graph) in which the pair of arcs $(u, v), (v, u) \in A(N)$ of capacity 1 corresponds to edge $e = \{u, v\} \in E(G)$;

vertices $v^+ \in V_{out}$, i.e. possible end-vertices of chains, are the sources of the unit flow, vertices $v^- \in V_{in}$, i.e. possible pierce points, are stocks.

Step 2. For network $N(V, A)$ get the circulation $x : A \to \{0,1\}$, i.e. the solution of the following problem

$$f(x) = \sum_{(u,v) \in A(N)} x(u,v) \to \min_x,$$

$$\sum_{v:\, (u,v) \in A(N)} x(u,v) - \sum_{v:\, (v,u) \in A(N)} x(v,u) = 1, \qquad u \in V_{out},$$

$$\sum_{v:\, (u,v) \in A(N)} x(u,v) - \sum_{v:\, (v,u) \in A(N)} x(v,u) \geq -1, \qquad u \in V_{in},$$

$$\sum_{v:\, (u,v) \in A(N)} x(u,v) - \sum_{v:\, (v,u) \in A(N)} x(v,u) = 0, \quad u \in V \backslash (V_{out} \cup V_{in}),$$

$$0 \leq x(u,v) \leq 1, \quad (u,v) \in A(N).$$

Circulation $x : A \to \{0,1\}$ can be obtained by solving the maximal flow of the minimal cost problem for a bipolar network. This network may be obtained by introducing a common source s adjacent to all sources and common source adjacent to all possible stocks in network $N(V, A)$. Moreover, the permissible circulation of $x : A \to \{0,1\}$ corresponds to the maximum flow value not exceeding $|V_{out}|$.

Step 3. If $f(x) < |V_{out}|$ then according to Proposition 3 $PPOE$-cover does not exist, go to Step 10.

Step 4. For each active arc $(u,v) : x(u,v) = 1$ create a list of arcs including it and only it.

Step 5. For each vertex $v \in V(G)$ "correctly" split vertices with the "correct" union of active edges lists (i.e. taking into account the cyclic order on the set of arcs and their orientation) while it is possible. The example of these "correct" splitting and union are shown in Fig. 4. As the result of this step we get the sequence of disjoint chains $C_1, C_2, \ldots C_{|V_{out}|}$ containing all flow holders and only them.

Step 6. Chande the arcs orientation for the obtained chains $C_1, C_2, \ldots C_{|V_{out}|}$.

Step 7. Continue each chain of the sequence $C_1, C_2, \ldots C_{|V_{out}|}$ while we get the maximal possible OE-chain. The result of this step is the initial part of the OE-cover $\tilde{C}_1, \tilde{C}_2, \ldots \tilde{C}_{|V_{out}|}$.

Step 8. Construct the partial graph

$$\tilde{G} = G \backslash \left(\bigcup_{i=1}^{|V_{out}|} \tilde{C}_i \right), \ E(\tilde{G}) = \left(E(G) \backslash \left(\bigcup_{i=1}^{|V_{out}|} \tilde{C}_i \right) \right),$$

in which all the vertices $v \in V_{out}$ avoiding piercing are the even ones. The lists of edges obtained as a result of the algorithm execution correspond to valid fragments of the cutting route.

Step 9. Define the shortest matching M on set $V_{odd}(\tilde{G})$. Run algorithm M-Cover [15] for $\tilde{G}$. We get chains $C_{|V_{out}|+1}, \ldots, C_{|V_{out}|+|M|}$.

Step 10. Stop.

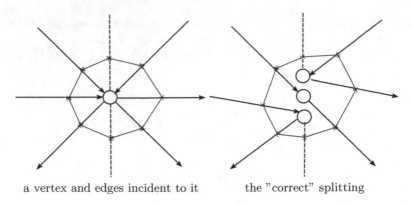

a vertex and edges incident to it   the "correct" splitting

Fig. 4. The example of the "correct" splitting

Theorem 1. *Let $G(V, E)$ be a plane graph without bridges.*
Algorithm PPOE-routing correctly solves the problem of obtaining the PPOE-cover for $G(V, E)$ by time not exceeding $O(|V|^3)$.

Proof. Let us remind that the main characteristic feature of algorithm M-Cover [15] is that unlike algorithm OE-Router, the next vertex $u = M(v) \in V_{Odd}$ is given for each $v \in V_{Odd}$. Algorithm M-Cover can interrupt the current chain both at the first visit of the vertex $v \in V_{Odd}$, and after this vertex becomes dead-end vertex (i.e. $Q(v) = \emptyset$).

$$\mathrm{Idle}_M(v) = (\mathrm{rank}(v) \le \mathrm{rank}(M(v))) \wedge \left(f_{M(v)} \succeq f_v \right), \quad v \in V_{Odd},$$

$$\text{where} \quad f_w = \arg \min_{f:v \in f \subset F(G)} \mathrm{rank}(f), \quad w \in V_{Odd}$$

are introduced. Partial order relation $\succeq$ is defined due to next

Remark 1. [15] If G be a plane connected graph on S without bridges than for any set M being a matching on set V_{odd} (vertices of odd degree) of graph G so that $(M \cap S) \backslash V = \emptyset$, there exists such an Eulerian cycle $C = v_1 e_1 v_2 e_2 ... e_n v_1$, $n = |E| + |M|$ for any initial part $C_l = v_1 e_1 v_2 e_2 ... v_l$, $l \le |E| + |M|$ of which the condition $\mathsf{Int}(C_l) \cap G = \emptyset$ holds.

The route consisting of chains $\tilde{C}_1, \tilde{C}_2, ... \tilde{C}_{|V_{out}|}$ is the edge-disjoint $PPOE$-route (due to the splitting and application of appropriate algorithms). Partial graph $\tilde{G}$ does not contain any edges belonging to chains $\tilde{C}_i$, $i = 1, ..., |V_{out}|$ by definition. All graph $\tilde{G}$ vertices avoiding piercing have even degree due to constructions. As a result of running Step 9 we get the continuation

$$C_{|V_{out}|+1}, ..., C_{|V_{out}|+|M|}$$

of route which is the OE-route in graph $\tilde{G}$ covering all edges of graph $\tilde{G}$, and vertex $v \in V_{in}$ for each chain C_i, $i = |V_{out}| + 1, ... |V_{out}| + |M|$ is permissible for piercing. Hence, the route

$$\tilde{C}_1, \tilde{C}_2, \ldots \tilde{C}_{|V_{out}|}, C_{|V_{out}|+1}, \ldots, C_{|V_{out}|+|M|}$$

is $PPOE$-cover of initial graph G.

Let us estimate the computing complexity of this algorithm. Step 1 allows to get the network by time $O(|E|)$. Circulation at step 2 may be obtained by time not exceeding $O(|V|^3)$ [8]. Step 3 is to verify the condition and is completed in $O(1)$. At step 4, a plurality of chains along a plurality of active arcs is introduced. This operation is performed in a time not exceeding $O(|E|)$. In step 5, at each vertex v, a "merging" of lists is performed in a time not exceeding $O(|V| \cdot \deg(v))$. Thus, computing complexity of step 5 does not exceed the value $O(|V| \cdot |E|)$. Step 6 runs the time not exceeding $O(|E|)$. The complexity of Step 7 is defined by complexity of algorithm OE-Cover [15] and amounts to $O(|E(G)| \cdot \log_2 |V(G)|)$. Obtaining the partial graph $\tilde{G}$ at Step 8 claims the time not exceeding $O(|E|)$. The complexity of Step 9 does not exceed $O(|V|^3)$ used for the shortest matching obtaining. Thus, the complexity of algorithm $PPOE$-routing does not exceed the value of $O(|V|^3)$.

So, the construction of the $PPOE$-cover of the G graph allows us to solve the problems of the cutter movement routing for an realizable cutting plan with restrictions on possible puncture points.

5 Conclusions

The problem of technological preparation of cutting processes has a pronounced systemic character, since it concerns both the problem of drawing up cutting plans and the problem of routing the cutting tool. The cutting plan must meet the following criteria: minimization of unproductive material consumption, realizability of the cutting plan on the cutting device, minimization of the cut length, minimization of the length of idle transitions. The methods of routing solutions presented in this paper allow us to determine the feasibility of the cutting plan on cutting machines, determine the length of the cut, as well as the number and length of idle transitions. Thus, the results of this research allow us to give a systematic estimation of realizability of the given cutting plan and solve the problem of finding the acceptable optimal trajectory of the cutter.

The subject of the further research is the software development for $PPOE$-chains constructing algorithm and development of the library of classes for routing problems using plane graphs solution.

References

1. Adelmann, B., Hellmann, R.: Fast laser cutting optimization algorithm. Phys. Proc. **12**, 591–598 (2011). https://doi.org/10.1016/j.phpro.2011.03.075
2. Chentsov, A., Khachay, M., Khachay, D.: Linear time algorithm for precedence constrained asymmetric generalized traveling salesman problem. In: 8th IFAC Conference on Manufacturing Modelling, Management and Control MIM 2016, Troyes, France, 28–30 June 2016, vol. 49, pp. 651–655. IFAC-PapersOnLine. https://doi.org/10.1016/j.ifacol.2016.07.767

3. Chentsov, A.G., Grigoryev, A.M., Chentsov, A.A.: Solving a routing problem with the aid of an independent computations scheme. Vestnik Yuzhno-Ural'skogo gosudarstvennogo universiteta. Seriya: Vychislitel'naya matematika i informatika [Bull. South Ural State Univ. Ser.: Comput. Math. Inform.] **11**(1), 60–74 (2018). https://doi.org/10.14529/mmp180106

4. Dewil, R., Vansteenwegen, P., Cattrysse, D., Laguna, M., Vossen, T.: An improvement heuristic framework for the laser cutting tool path problem. Int. J. Prod. Res. **53**(6), 1761–1776 (2015). https://doi.org/10.1080/00207543.2014.959268

5. Dewil, R., Vansteenwegen, P., Cattrysse, D.: Construction heuristics for generating tool paths for laser cutters. Int. J. Prod. Res. **52** (2014). https://doi.org/10.1080/00207543.2014.895064

6. Dewil, R., Vansteenwegen, P., Cattrysse, D.: A review of cutting path algorithms for laser cutters. Int. J. Adv. Manuf. Technol. **87**, 1865–1884 (2016). https://doi.org/10.1007/s00170-016-8609-1

7. Gadallah, M.H., Abdu, H.M.: Modeling and optimization of laser cutting operations. Manuf. Rev. **2**, 1–15 (2015). https://doi.org/10.1051/mfreview/2015020

8. Papadimitriou, C.H., Steiglitz, K.: Combinatorial Optimization: Algorithms and Complexity. Prentice Hall, Englewood Cliffs (1982)

9. Hajad, M., Tangwarodomnukun, V., Jaturanonda, C., Dumkum, C.: Laser cutting path optimization using simulated annealing with an adaptive large neighborhood search. Int. J. Adv. Manuf. Technol. **103**, 781–792 (2019). https://doi.org/10.1007/s00170-019-03569-6. https://www.springerprofessional.de/en/the-international-journal-of-advanced-manufacturing-technology/4883186

10. Khachay, M., Neznakhina, K.: Towards tractability of the Euclidean generalized traveling salesman problem in grid clusters defined by a grid of bounded height. In: Eremeev, A., Khachay, M., Kochetov, Y., Pardalos, P. (eds.) OPTA 2018. CCIS, vol. 871, pp. 68–77. Springer, Cham (2018). https://doi.org/10.1007/978-3-319-93800-4_6

11. Lee, M.K., Kwon, K.B.: Cutting path optimization in CNC cutting processes using a two-step genetic algorithm. Int. J. Prod. Res. **44**(24), 5307–5326 (2006). https://doi.org/10.1080/00207540600579615

12. Makarovskikh, T.: The estimation of OE-cover cardinality for a plane graph. Bull. USATU **21**(2(76)), 112–118 (2017). (in Russian)

13. Makarovskikh, T., Panyukov, A.: Development of routing methods for cutting out details. In: Belin, S., Kononov, A., Kovalenko, Y. (eds.) CEUR Workshop Proceedings, vol. 2098, pp. 249–263. Dostoevsky Omsk State University, Omsk, Russia, July 2018. http://ceur-ws.org/Vol-2098

14. Makarovskikh, T., Panyukov, A., Savitskiy, E.: Mathematical models and routing algorithms for cad technological preparation of cutting processes. Autom. Remote Control **78**(4), 868–882 (2017). https://doi.org/10.14529/mmp180104

15. Makarovskikh, T., Panyukov, A., Savitskiy, E.: Mathematical models and routing algorithms for economical cutting tool paths. Int. J. Prod. Res. **56**(3), 1171–1188 (2018). https://doi.org/10.1080/00207543.2017.1401746

16. Makarovskikh, T., Savitskiy, E.: The OE-cover for a plane graph by chains with allowed starting vertices. In: CEUR Workshop Proceedings, vol. 2064, pp. 103–111 (2017). http://ceur-ws.org/Vol-2064/paper12.pdf

17. Manber, U., Israni, S.: Pierce point minimization and optimal torch path determination in flame cutting. J. Manuf. Syst. **3**(1), 81–89 (1984). https://doi.org/10.1016/0278-6125(84)90024-4. http://www.sciencedirect.com/science/article/pii/0278612584900244

18. Panyukova, T.: Chain sequences with ordered enclosing. J. Comput. Syst. Sci. Int. **46**(1), 83–92 (2007). https://doi.org/10.1134/S1064230707010108

19. Panyukova, T., Panyukov, A.: The algorithm for tracing of flat Euler cycles with ordered enclosing. Bull. Chelyabinsk Sci. Center Ural Branch Russ. Acad. Sci. **4**(9), 18–22 (2000). https://doi.org/10.1134/S1064230707010108. https://link.springer.com/article/10.1134/S1064230707010108

20. Panyukova, T.: Constructing of OE-postman path for a planar graph. Bull. South Ural State Univ. Ser.: Math. Model. Program. Comput. Softw. **7**(4), 90–101 (2014)

21. Petunin, A.A., Polishchuk, E.G., Ukolov, S.S.: On the new algorithm for solving continuous cutting problem. IFAC-PapersOnLine **52**(13), 2320–2325 (2019). https://doi.org/10.1016/j.ifacol.2019.11.552

22. Petunin, A., Stylios, C.: Optimization models of tool path problem for CNC sheet metal cutting machines. IFAC-PapersOnLine **49**, 23–28 (2016)

23. Silva, E.F., Oliveira, L.T., Oliveira, J.F., Toledo, F.M.B.: Exact approaches for the cutting path determination problem. Comput. Oper. Res. **112**, 104772 (2019). https://doi.org/10.1016/j.cor.2019.104772. https://doi.org/10.1016/j.cor.2019.104772Getrightsandcontent

Effectiveness of Nash Equilibrium Search Algorithms in Four-Person Games in General and Multi-matrix Settings

Ustav Malkov[1]([envelope]) [iD] and Vlasta Malkova[2] [iD]

[1] Central Economics and Mathematics Institute of Russian Academy of Science, Nakhimovsky Prospect 32, 117418 Moscow, Russia
ustav-malkov@yandex.ru
[2] Dorodnicyn Computing Centre, FRC CSC RAS, Vavilov St. 40, 119333 Moscow, Russia
vmalkova@yandex.ru

Abstract. The effectiveness of using the local search algorithm (mountain climbing) and the Lemke–Howson method for searching for Nash equilibrium in 4-person games in general and multi-matrix settings using the Matlab, Python, and FORTRAN software environments are studied. The local search procedure implemented in the Python environment, involving the use of multiplication of a multi-dimensional matrix by a vector, turned out to be an effective tool. The modification of the Lemke–Howson method for multi-matrix formulation showed very good results.

Keywords: Bimatrix · Game · Multi-matrix game · General setting · Mixed strategies · Nash equilibrium · Lemke–Howson algorithm · Python · Matlab

1 Introduction

As we know, the search for Nash equilibrium in several persons games can be considered as a non-linear programming problem, which by fixing strategies all but one player is turned into a linear programming problem. Solving these problems consequently we get the local search algorithm (LS). This algorithm called "mountain climbing" was proposed in [5] and has been successfully applied for searching the Nash equilibrium in bimatrix games [6]. The same approach (LS) is applicable for three-person games in both general [3,4] and multi-matrix [8] settings. The Lemke–Howson method (LH-algorithm) has performed well in the case of bimatrix games [7], but it is not applicable for three-person games in a general setting. However, it turned out that the modification [4] of the LH-algorithm works well for three-person games in a multi-matrix setting. In this paper, we investigate the effectiveness of the considered approaches for Nash equilibrium

The work was supported by the Russian Foundation for Basic Research (project No. 20-010-00169a).

N. Olenev et al. (Eds.): OPTIMA 2020, CCIS 1340, pp. 198–208, 2020.
https://doi.org/10.1007/978-3-030-65739-0_15

search for 4-person games in general and multi-matrix settings. When switching from 3-person games to 4-person games, the amount of information and the search time when using the local search algorithm, it increases by more than an order of magnitude. Therefore, if it was possible to solve 3-persons games up to dimension 100 (i.e. up to 100 strategies per player), then in the case of games 4 persons can only get a solution within a reasonable time for several dozen strategies.

Earlier in the presentation of the LS-algorithm for 3-person games [3] we represented the iteration of the algorithm as a sequential formation and solution of three LP problems. In this paper, for 4-person games, we use a compact description of the LS-algorithm, where 4 LP problems are formed and are solved in one cycle. This representation of the algorithm made it possible to compactly write and implement codes in Matlab and Python environments. It became much easier to debug programs by making changes in one place of the text, not in 4.

The multi-matrix formulation we use is a generalization for 4-person games hexamatric setting for 3-person games, where the tables of players' winnings are defined by 12 matrices. In this case, the n-person games as nonlinear programming problems becomes a quadratic programming ones with linear conditions and this circumstance adds the ability to solve 4-person games using a modification of the Lemke–Howson algorithm, which showed good results when solving 3-person games. Performed numerical experiments show that for 4-person games using this modification one can also successfully solve games with dimensions up to several hundred strategies.

There are several other approaches to solve 4-person games [1,2], including the multi-matrix (triplo) setting, where the payoff tables of players are defined by 12 three-dimensional tables. These approaches use a global optimization algorithm.

2   Four-Person Game in General Setting

Consider the game Γ with four players, i.e. $n = 4$. A finite non-cooperative 4-person game Γ is defined by four sets X_1, X_2, X_3, X_4 of strategies of the first, second, third and fourth player respectively, where $1 \leqslant r \leqslant 4$

$$X_r = \{x_r = (x_1^r, \ldots, x_{m_r}^r) \in \mathbf{E}^{m_r} : \quad \langle x_r, e_{m_r} \rangle = 1, \quad x_r \geqslant o_{m_r}\}, \quad (1)$$

together with their payoff functions as follows

$$f^r(x) = f^r(x_1, x_2, x_3, x_4) = \sum_{i=1}^{m_1} \sum_{j=1}^{m_2} \sum_{k=1}^{m_3} \sum_{l=1}^{m_4} A_{ijkl}^r \, x_i^1 \, x_j^2 \, x_k^3 \, x_l^4, \quad (2)$$

$$f(x_1, x_2, x_3, x_4) = \sum_{i=1}^{m_1} \sum_{j=1}^{m_2} \sum_{k=1}^{m_3} \sum_{l=1}^{m_4} A_{ijkl} \, x_i^1 \, x_j^2 \, x_k^3 \, x_l^4, \quad (3)$$

where A_{ijkl}^r are the players' 4-dimensional payoff tables,

$$A_{ijkl} = \sum_{r \in I} A_{ijkl}^r, \quad I = \{1, 2, 3, 4\}.$$

Strategies x_1, x_2, x_3, x_4 have dimensions m_1, m_2, m_3 and m_4, vector $x = (x_1, x_2, x_3, x_4) \in \mathbf{E}^M$, $M = m_1 + m_2 + m_3 + m_4$.

By entering the designation $\{X_1, X_2, X_3, X_4\} = \{X, Y, Z, W\}$, $\{A^1, A^2, A^3, A^4\} = \{A, B, C, D\}$, $\{i, j, k, l\} = \{j_1, j_2, j_3, j_4\}$, one can get formulas that look more like formulas [4] for three-players game ($n = 3$).

In the future without reducing the generality for ease of presentation we define $m_1 = m_2 = m_3 = m_4 = m$, $M = 4\,m$.

The Nash indicator for game (1)–(3) is

$$N(x_1, x_2, x_3, x_4) = \max_{x_1' \in X_1} f_1(x_1', x_2, x_3, x_4) + \max_{x_2' \in X_2} f_2(x_1, x_2', x_3, x_4)$$

$$+ \max_{x_3' \in X_3} f_3(x_1, x_2, x_3', x_4) + \max_{x_4' \in X_4} f_4(x_1, x_2, x_3, x_4') - f(x_1, x_2, x_3, x_4). \quad (4)$$

The local search for the minimum of the Nash indicator (4) turns into a sequential solution of 4 linear programming problems to determine their solutions x_i^*, $i \in I$ as mixed strategies of players. These problems are solved relative to the strategy of one of the players with fixed strategies of other players, which were obtained in previous iterations as optimal strategies of these players.

3   The Local Search Algorithm

As the initial (starting) strategy of an iterative process, it is usually a set of pure strategies is selected, for example, $x^0 = (\bar{x}_1, \ldots, \bar{x}_n)$, where $\bar{x}_i = (1, 0, \ldots, 0) \in \mathbf{E}^{m_i}$, $i \in I - 1$. So as a starting point, let define the strategies x^2, x^3, x^4 of players 2, 3 and 4 as their first pure strategies $\bar{x}^2 = (1, 0, 0 \ldots)$, $\bar{x}^3 = (1, 0, 0 \ldots)$, $\bar{x}^4 = (1, 0, 0 \ldots)$.

We solve four LP problems sequentially, one after the other until we get the local minimum of the Nash function. If this minimum is zero (or does not exceed the preset small positive number), then the equilibrium (approximate equilibrium) is found. Otherwise, select a different starting point and repeat the procedure again.

By setting the ratio $\{p_1, p_2, p_3, p_4\} = \{i, j, k, l\}$ and $I - r = \{s_1, s_2, s_3\}$ (See below), using a single formula (6), it is possible to write 12 systems of inequalities for four LP problems and one operator in the program text of implementations in MATLAB and Python environments. And, accordingly, collect and solve four LP problems in a cycle.

Earlier in the presentation of the LS-algorithm for 3-person games, we represented the iteration of the algorithm as a sequential formation and solution of three LP tasks. In this paper, for 4-person games, we use a compact description of the LS-algorithm, where 4 LP problems are formed and solved in one cycle. This representation of the algorithm made it possible to compactly write and implement codes in Matlab and Python environments.

Let solve four LP problem in compact description:
for $r \in I = (1, 2, 3, 4)$ (denoting the set $I - r = \{s_1, s_2, s_3\}$).

and form the functional of the problem r

$$\sum_{p_r=1}^{m} \left(\sum_{s_1=1}^{m} \sum_{s_2=1}^{m} \sum_{s_3=1}^{m} A_{p_r p_{s_1} p_{s_2} p_{s_3}} \overline{x}_{p_{s_1}}^{s_1} \overline{x}_{p_{s_2}}^{s_2} \overline{x}_{p_{s_3}}^{s_3} \right) x_{p_r}^r - \sum_{u \in I-r} \alpha_u^r \to \max_{x_{p_r}^r, \alpha_u^r}, \quad (5)$$

and the system of m inequalities (one of the three systems of the problem r): for $u \in I - r$ (taking the notation $I - u - r = \{t_1, t_2\}$, so $v = u = I - r - t_1 - t_2$, and a correspondence $\{i, j, k, l\} = \{p_v, p_{t_1}, p_{t_2}, p_r\}$ $p = \{i, j, k, l\}$)

$$\sum_{p_r=1}^{m} \left(\sum_{p_{t_1}=1}^{m} \sum_{p_{t_2}=1}^{m} A_{p_v \, p_{t_1} \, p_{t_2} \, p_r}^u \overline{x}_{p_{t_1}}^{t_1} \overline{x}_{p_{t_2}}^{t_2} \right) x_{p_r}^r \leqslant \alpha_u^r,$$
$$\sum_{p_r=1}^{m} x_{p_r}^r = 1, \quad x_{p_r}^r \geqslant o_m, \quad p_v \in \{1, \dots, m\}, \quad (6)$$

or denoting $H_{p_v, p_r} := \sum_{p_{t_1}=1}^{m} \sum_{p_{t_2}=1}^{m} A_{p_v \, p_{t_1} \, p_{t_2} \, p_r}^u \overline{x}_{p_{t_1}}^{t_1} \overline{x}_{p_{t_2}}^{t_2}$, we get

$$\sum_{p_r=1}^{m} H_{p_v, p_r} x_{p_r}^r \leqslant \alpha_u^r, \quad \sum_{p_r=1}^{m} x_{p_r}^r = 1, \quad x_{p_r}^r \geqslant o_m, \quad p_v \in \{1, \dots, m\}. \quad (7)$$

The implementation of the LS-algorithm in the Python environment has two modes. In mode 1, the LP problem $\mathcal{A}$ matrix is formed with nested loops using four indexes $\{i, j, k, l\} = \{p_1, p_2, p_3, p_4\}$. In mode 0, only two indexes i and l. In this case, the inner loop uses the operation of multiplying the four-dimensional table (this operation is in Python, but not in Matlab) by two vectors in turn. In the first mode, a cycle of four indexes is a long operation and for $m = 100$ takes an unacceptably long time ≈ 2000 s. In the second case, using matrix multiplication by a vector reduces the time dramatically by two orders of magnitude (for $m = 100$, it is about 12 s instead of 2000 s).

Let's solve the LP problem constructed in this way and replace the previously fixed strategy $\overline{x}_r$ for player r with the obtained solution x_r^*.

4   Multi-matrix Setting

The multi-matrix formulation we use is a generalization for 4-person games hexa-matric setting for 3-person games, where the tables of players' winnings are defined by 12 matrices. In a multi-matrix setting, playoff functions of n players are set by the formula

$$f^r(x) = f^r(x_1, \dots, x_n) = \left\langle x_r, \sum_{q=1}^{n} A_{rq}^q x_q \right\rangle, \quad 1 \leqslant r \leqslant n, \quad (8)$$

where A_{rq}^r are $m_r \times m_q$ matrices, $r, q \in \{1, \dots, n\}$, A_{rr}^r are zero matrices, $1 \leqslant r \leqslant n$.

For a 4-person game ($n = 4$), we use formulas similar to the case of hexam-atric games [8] ($n = 3$):

$$f_x(x, y, z, w) = \langle x, a_{ij}^1 \, y + a_{ik}^2 \, z + a_{il}^3 \, w \rangle,$$
$$f_y(x, y, z, w) = \langle y, b_{ji}^1 \, x + b_{jk}^2 \, z + b_{jl}^3 \, w \rangle,$$
$$f_z(x, y, z, w) = \langle z, c_{ki}^1 \, x + c_{kj}^2 \, y + c_{kl}^3 \, w \rangle, \qquad (9)$$
$$f_w(x, y, z, w) = \langle w, d_{li}^1 \, x + d_{lj}^2 \, y + d_{lk}^3 \, z \rangle;$$

and the players' payoff tables are represented using the following formulas:

$$A_{ijkl} = a_{ij}^1 + a_{ik}^2 + a_{il}^3, \qquad B_{ijkl} = b_{ji}^1 + b_{jk}^2 + b_{jl}^3,$$
$$C_{ijkl} = c_{ki}^1 + c_{kj}^2 + c_{kl}^3, \qquad D_{ijkl} = d_{li}^1 + d_{lj}^2 + d_{lk}^3.$$

For a compact representation of the local search algorithm, we gathered these matrices in a block matrix

$$H = \begin{pmatrix} O_{m_1} & a_{ij}^1 & a_{ik}^2 & a_{il}^3 \\ b_{ji}^1 & O_{m_2} & b_{jk}^2 & b_{jl}^3 \\ c_{ki}^1 & c_{kj}^2 & O_{m_3} & c_{kl}^3 \\ d_{li}^1 & d_{lj}^2 & d_{lk}^3 & O_{m_4}, \end{pmatrix}$$

where the matrices $H_{i,j}$ are the components of this block matrix, for example, $H_{1,2} = a_{ij}^1$.

The algorithm for local search of Nash equilibrium for the 4-person game in the multi-mastrix setting takes the following form.

Let $\{x, y, z, w\} = \{x_1, x_2, x_3, x_4\}$. In the loop, we will collect the functionals and condition matrices of four linear programming problems: for $r \in I$ and $\{s, t, v\} = I - r$, the criterion of the problem r is

$$\left\langle x_r, \sum_{v \in I-r} (H_{r,v} + H_{v,r}^{\mathrm{T}}) \bar{x}_v \right\rangle - \sum_{v \in I-r} \alpha_v^r \to \max_{x_r, \alpha_q^r}; \qquad (10)$$

and the system of inequalities (one of the three systems of the problem r)

$$\langle H_{u,r}^{\mathrm{T}} \bar{x}_u, x_r \rangle + \langle \bar{x}_u, H_{u,s} \bar{x}_s + H_{u,t} \bar{x}_t \rangle \leqslant \alpha_u^r,$$
$$\langle x_r, e_m \rangle = 1, \quad x_r \geqslant o_m, \quad \alpha_u^r \geqslant 0_1, \quad u \in I - r. \qquad (11)$$

In this case, the n-person game as nonlinear programming problem become quadratic programming ones with linear conditions and this circumstance adds the ability to solving 4-person games using a modification of the Lemke–Howson algorithm, which showed good results when solving 3-person games. Conducted numerical experiments show that for 4-person games using this modification one can also successfully solve games with dimensions up to several hundred strategies.

5 Modification of the Lemke–Howson Algorithm for Finding the Nash Equilibrium in the 4-Person Multi-matrix Game

The Lemke–Howson algorithm looks for a solution of the system of equations $AX + Y = e$ ($e = (1, \ldots, 1)$) for which the complementarity conditions

$(X, Y) = 0$ are satisfied. The algorithm is universal and works the same way for different games, both for a bimatric game, and for auxiliary problems of three and four persons games. For all these cases it is necessary to collect the corresponding matrices A from the matrices of the players payoff tables

$$A_{2m,2m} = \begin{pmatrix} 0 & A_1 \\ B_1 & 0 \end{pmatrix}, \quad A_{3m,3m} = \begin{pmatrix} 0 & A_1 & A_2 \\ B_1 & 0 & B_2 \\ C_1 & C_2 & 0 \end{pmatrix}, \quad A_{4m,4m} = \begin{pmatrix} 0 & A_1 & A_2 & A_3 \\ B_1 & 0 & B_2 & B_3 \\ C_1 & C_2 & 0 & C_3 \\ D_1 & D_2 & D_3 & 0 \end{pmatrix}.$$

(12)

The LH-algorithm is applicable to the auxiliary problem that we form in the form of the following system of linear equations:

$$A X + Y = e, \quad X = (x_1, \ldots, x_{4m}), \quad Y = (y_1, \ldots, y_{4m}),$$

$A = A_{4m,4m}$ is table of winnings, X is vector of player strategies, Y is vector of additional variables, $e = (1, \ldots, 1) \in \mathbf{E}^{4m}$.

The matrix $A_{4m,4m}$ is obtained by converting the multi-dimensional table $H_{m,m,4,4}$ of player winnings. The solution to the auxiliary problem is found by referring to the LH procedure (LH-algorithm). The Lemke–Howson algorithm consists of the following steps. First, we form the initial basis from additional variables Y. Next, we enter one of the structural variables x into the basis, for example, x_1. This violates one of the complementarity conditions $x_1 * y_1 > 0$. Variables of this system of equations form pairs x_i, y_i according to complementarity conditions, and in the basis from each pair can be only one. If when we enter the next variable into the basis, some variable comes out of the basis that is defined uniquely in the case of undegenerate systems of equations, then one can enter its partner into the basis, without violating additional complementary condition. By entering the variables in base this way, without increasing the number of violated conditions, until one of the variables from the pair x_1, y_1 leaves the base, thus reaching an equilibrium with the fulfillment of the complementarity conditions. Since the variables derived from the basis and their partners are uniquely defined, the sequence (Lemke's path) of passing the vertices of the polyhedron of the system of conditions is determined by the variable entered in the basis first.

The solution of the auxiliary problem is found by referring to the procedure LH (Lemke–Howson algorithm) using the formula: $[x, y, z, w] = LH(m, A)$. Let use notations $((x, y, z, w) = X_{4m} = X_{4,m})$.

If we choose the following matrices $A_1, A_2, A_3, B_1, B_2, B_3, C_1, C_2, C_3, D_1, D_2, D_3$ as the ones that form the payoff tables, then this problem looks as follows

$$
\begin{aligned}
A_1 y + A_2 z + A_3 w + u &= \alpha e_m, & B_1 x + B_2 z + B_3 w + v &= \beta e_n, \\
C_1 x + C_2 y + C_3 w + f &= \lambda e_k, & D_1 x + D_2 y + D_3 z + g &= \mu e_l, \\
\langle x, e_m \rangle = 1, & \quad x \geqslant 0_m, & \langle y, e_n \rangle = 1, & \quad y \geqslant 0_n, \\
\langle z, e_k \rangle = 1, & \quad z \geqslant 0_k, & \langle w, e_k \rangle = 1, & \quad w \geqslant 0_l, \\
x, u \geqslant 0, \quad y, v \geqslant 0, & \quad z, f \geqslant 0, & w, g \geqslant 0, & \quad \alpha, \beta, \lambda, \mu \geqslant 0, \\
\langle x, u \rangle = 0, & \quad \langle y, v \rangle = 0, & \langle z, f \rangle = 0, & \quad \langle w, g \rangle = 0.
\end{aligned}
$$

(13)

Due to the linearity of the conditions of the problem Γ for a four-person game in a multi-matrix setting we would like to apply the Lemke–Howson algorithm to the game Γ, as it is done for bimatrix games, where replacing the variables $x' = x/\alpha, y' = y/\beta$ translates the original problem to an equivalent linear complementarity one. For the four persons game, this technique does not work. However, having solved a series of auxiliary problems, which we obtain defining $\alpha = \beta = \gamma = \mu = 1$, using the LH-algorithm, we can find the Nash equilibrium point.

For the solution of the game Γ, the complementarity conditions

$$x_i u_i = 0, \quad 1 \leqslant i \leqslant m, \quad y_j v_j = 0, \quad 1 \leqslant j \leqslant m,$$
$$z_k f_k = 0, \quad 1 \leqslant k \leqslant m, \quad w_k g_l = 0, \quad 1 \leqslant l \leqslant m$$

must be fulfilled.

Defining $\alpha = \beta = \gamma = \mu = 1$, we obtain an auxiliary problem in the form of a system of equations

$$A_1 y + A_2 z + A_3 w + u = e, \quad B_1 x + B_2 z + B_3 w + v = e,$$
$$C_1 x + C_2 y + C_3 w + f = e, \quad D_1 x + D_2 y + D_3 z + g = e,$$
$$\langle x, u \rangle = 0, \quad \langle y, v \rangle = 0, \quad \langle z, f \rangle = 0, \quad \langle w, g \rangle = 0,$$
$$x, u \geqslant 0, \quad y, v \geqslant 0, \quad z, f \geqslant 0, \quad w, g \geqslant 0.$$

Enter the notation

$$H_{4m,4m} = \begin{pmatrix} 0 & A_1 & A_2 & A_3 \\ B_1 & 0 & B_2 & B_3 \\ C_1 & C_2 & 0 & C_3 \\ D_1 & D_2 & D_3 & 0 \end{pmatrix}, s = (s_1, ..., s_p)^{\mathrm{T}} = \begin{pmatrix} x \\ y \\ z \\ w \end{pmatrix}, \sigma = (\sigma_1, ..., \sigma_p)^{\mathrm{T}} = \begin{pmatrix} u \\ v \\ f \\ g \end{pmatrix}.$$

In these notations, the linear complementarity problem is finding a non-negative (s, σ) solution for the system $Hs + E\sigma = e$, under complementarity conditions $(s_i, \sigma_i) = 0, i = 1, \ldots, p$. Here $e = e_p$, $E = \text{diag}(e)$, and E is a unit matrix of size p.

Let's apply to this system the procedure Lemke–Howson. We take σ as the initial basis. At first, we enter the variable s_1 into the basis. Some variable comes out of the basis that was selected from the condition not to violate the nonnegativity of the basic variables. Next, we enter a variable in the basis that is associated with complementarity conditions with the variable that came out of the basis. Let's repeat this operation until we exclude from the basis the variable s_1.

As the result, we get a "pseudo-Nash equilibrium". Since the Lemke–Howson method can loop, starting from the next starting point, we perform a specified (limited) number of iterations, for example, p (the dimension of the game). Here, the "pseudo-Nash equilibrium" is taken in quotation marks, since the conditions $x^{\mathrm{T}} e_m = 1$, $y^{\mathrm{T}} e_n = 1$, $z^{\mathrm{T}} e_l = 1$, and $w^{\mathrm{T}} e_q = 1$ were not taken into account when solving the auxiliary system.

Starting points are selected one by one from the set $\{x_1, \ldots, x_m, y_1, \ldots, y_n, z_1, \ldots, z_l, w_1, \ldots, w_q\}$. We use the resulting basis of auxiliary problem as the initial basis for the four-persons game with all conditions and variables $(\alpha, \beta, \gamma \, and \, \mu)$. This initial basis is constructed as follows. First, we enter variables from the auxiliary task basis into the basis in the positions where they were in the auxiliary task basis. There are probably still some variables that can't be entered in their positions, since the leading positions of the corresponding columns decomposing by the current basis may have very small absolute values. These variables are entered into the basis in the positions that are not still occupied, where there are coefficients in the leading positions that are sufficiently large in absolute value. The resulting solution may contain negative values. In this case, we will repeat the solution of the auxiliary problem from the next starting point. The resulting point with non-negative values of the basic variables has the same structure as the solution of the auxiliary problem, that is, the complementarity conditions are met for it, and it is a Nash equilibrium point.

As shown by the solution of test games, it was enough to iterate over a relatively small number of "pseudo-Nash equilibria" to get a result. If one can build a basis for the full system conditions, achieved using a basis when solving the auxiliary problem, and thus obtain the solution of the complete system is valid, then the game is solved. I.e. we get the point where the complementary conditions are fulfilled, the conditions of Nash equilibrium. Otherwise, we enter the following variable X in the initial basis of the auxiliary problem and repeat these actions until we get the result.

6   Testing the Local Search Algorithm and the Lemke–Howson Method

LS local search and LH Lemke–Howson algorithms for games with $n = 4$ players are implemented in Matlab, Python, and Fortran software environments. All test issues were generated using a random number of sensors that are different for different environments. Therefore, the tested games are different in different environments, and it would be necessary to perform calculations for at least 20 games of the same dimension, and then average the results. However, due to the large time spent, we were forced to limit ourselves to 5 games. First, we found out the speed of these implementations on an example of the game $100 \times 100 \times 100 \times 100$, performing one iteration (building and solving four linear programming problems consequentially), to select the most effective implementations and execution modes of the LS-algorithm during testing. In the case of a general statement, it is necessary to collect 12 matrices (for 4 LP problems) of dimension $m \times m$, performing nested loops with the indices i, j, k, and l. For $m = 100$, we will have to perform about $100,000,000(100^4)$ operations. This means that implementations in MatLab and Python will have a long-running time (mode = 1). In the Python implementation (mode = 0), instead of loops with indexes i, j, k, and l, loops are performed only with two indexes, and with operations of multiplying a four-dimensional table by a vector, and the execution time will be completely different. In the case of a multi-matrix formulation, the matrices of

LP problems are collected from the matrices, components of payoff tables. And accordingly, there are no such deeply nested cycles (Table 1).

Table 1. The times in seconds of a single iteration of the game 100^4

Game 100^4	General setting				Multi-matrix setting
	Rezim 1		Rezim 0		
	Cycle	4LP	Cycle	4LP	
Mathlab	8775	3214	—	—	0,5
Python	3881	1160	21	8	0,5

Here, "cycle" means that 4 LP problems are formed and solved consequentially in a loop, and "4 LP" indicates that these operations are performed separately.

Table 2. Nash equilibrium search times in 5 games 25^4 by local search algorithm

	startp	*itn*	*time*	Settings	Regime
Mathlab	22456	213429	51942	Multi-matrix	
Python	21934	69068	3190		
	21697	68206	7723	General	4LP, regime = 0
	21934	72594	20916		Cycle, regime = 0
Fortran	212	1819	4757		

Table 2 shows the results obtained when solving a series of five games 25^4 using the LS-algorithm (and only one game in the Fortran environment due to large time's costs). Here *startp* – total number of used start points, *itn* – total number of steps of the algorithm, *time* – total time in seconds of the algorithm.

As can be seen from the results obtained, the LS-algorithm works effectively in the case of multi-matrix formulation, and in the case of General formulation, only with implementation in the Python environment in regime = 0 and 4LP mode, one can get results in a reasonable amount of time.

For a compact description of the LS-algorithm, forming matrices of 4 LP problems it is performed in a cycle, and this technique also allows we to compactly write program text of implementations. But at the same time, we have to perform additional actions in the innermost matrix formation cycle, compared to the case when these matrices are formed as separate operations. And so in cycle mode we have to spend significantly more time (Table 3).

Here "neg" is the number of obtained solutions of the auxiliary problem that were unfeasible for the initial game. This means that in the study of 5 games 25^4, 16 auxiliary problems were solved, and among them, in five cases from these

Table 3. Application of the LH-algorithm to games 25^4, 100^4 and 500^4.

Games	$startp$	itn	neg	$time$
5×25^4	111	2804	8	4,38
5×100^4	455	45524	14	106
1×500^4	1666	831566	6	28521

solutions, it was possible to construct a valid solution of the original game, i.e., get the Nash equilibrium. If in local search programs, the iteration is to construct and solution of 4 LP problems, then in the Lemke–Howson method modification algorithm, the iteration is one step of this method, i.e. one iteration of the simplex method algorithm over a matrix H of dimension $(4m) \times (4m)$. It is important to note that this matrix recalculation is not performed on all elements (of the order of $m \times n$ operations), but line by line (there is such an operation in Python), i.e. of the order of m operations. Note that in the local search algorithm, the possible number of starting points, i.e. the pure strategies of players y, z, and w, is m^3, and for a modification of the Lemke–Howson method, this number is $4m$, i.e. the total dimension of the strategies of four players. An amazing result! The best result of applying the local search algorithm in the case of games 25^4 was obtained using a Python program in 2458 s. At the same time, these problems were solved by modifying the Lemke–Howson method in 4.38 s!!! Three orders of magnitude faster.

7 Conclusion

As we know, the search for Nash equilibrium in several persons game can be considered as a non-linear programming problem, which when fixing strategies all but one player is turned into a linear programming problem. Solving these problems consequently we get the LS-algorithm that we used for 4-person games, more successfully in a multi-matrix formulation, and in the case of a General formulation using the implementation in Python, with procedures of multiplying a multi-dimensional table sequentially by two vectors. As a result, this implementation proved to be an effective tool for finding a Nash equilibrium. Out of competition was a modification of the LH-algorithm for multi-matrix formulation. A series of 5 games 25^4 in the multi-matrix formulation was able to solve by the LS-algorithm in 2458 s, and by modifying the LH-algorithm in 4.38 s! The Lemke–Howson method's modification for the multi-matrix formulation is out of the competition. There are several other approaches to solve 4-person games [1, 2], including the multi-matrix (triplo) setting, where the payoff tables of players are defined by 12 three-dimensional tables. These approaches use a global optimization algorithm. But compare the effectiveness of local search applied in this paper with the global search algorithm is not possible, because in the articles [1, 2] there is no data on the solution of a series of games, for example, 25^4.

There is only a confirmation that successfully they managed to solve several small games, including the game $6 \times 4 \times 3 \times 6$.

References

1. Batbileg, S., Tungalag, N., Anikin, A., Gornov, A., Finkelstein, E.: A global optimization algorithm for solving a four-person game. Optim. Lett. **13**(3), 587–596 (2017). https://doi.org/10.1007/s11590-017-1181-2
2. Enkhbat, R., Bathileg, S., Anikin, A., Tungalag, N., Gornov, A.: A note on four-players triple game. Contrib. Game Theory Manag. **12**, 100–112 (2019)
3. Golshtein, E.: A numerical method for solving finite three-person games. Economica i Matematicheskie Metody **50**(1), 110–116 (2014). (in Russian)
4. Golshtein, E., Malkov, U., Sokolov, N.: The Lemke-Howson algorithm solving finite non-cooperative three-person games in a special setting. DEStech Trans. Comput. Sci. Eng. (optim) 265–272 (2018). https://doi.org/10.12783/dtcse/optim2018/27938. (Supplementary volume)
5. Konno, H.: A cutting plane algorithm for solving bilinear programs. Math. Program. **11**(1), 14–27 (1976)
6. Lemke, C., Howson, C.: Equilibrium points of bimatrix games. J. Soc. Ind. Appl. Math. **12**(2), 413–423 (1964)
7. Orlov, A., Strekalovsky, A., Batbileg, S.: On computational search for nash equilibrium in hexamatrix games. Optim. Lett. **10**(2), 369–381 (2014)
8. Strekalovskii, A.S., Enkhbat, R.: Polymatrix games and optimization problems. Autom. Remote Control **75**(4), 632–645 (2014). https://doi.org/10.1134/S0005117914040043

Applications

Polynomially Solvable Subcases for the Approximate Solution of Multi-machine Scheduling Problems

Alexander Lazarev[1] , Darya Lemtyuzhnikova[1,2] , Nikolay Pravdivets[1(✉)] ,
and Frank Werner[3]

[1] Institute of Control Sciences, 65 Profsoyuznaya Street, 117997 Moscow, Russia
pravdivets@ipu.ru
[2] Moscow Aviation Institute, 4 Volokolamskoe Highway, 125993 Moscow, Russia
[3] Fakultät für Mathematik, Otto-von-Guericke-Universität Magdeburg,
PSF 4120, 39016 Magdeburg, Germany

Abstract. New metrics for different classes of scheduling problems are introduced. We show how approximate solutions of NP-hard problems can be obtained using these metrics. To do this, we solve the optimization problem in which the introduced metric is used as the objective function, and a system of linear inequalities of (pseudo-)polynomial solvable instances of the initial problem represents the constraints. As a result, we find a projection of the considered sub-instance onto the set of solvable cases of the problem in the introduced metric.

Keywords: Scheduling · Metric approach · Approximation

1 Introduction

A class of multi-machine scheduling problems is NP–hard in the strong sense so that the existence of a polynomial algorithm is unlikely. This has been proved for the special case $1|r_j|L_{max}$ of this problem with one machine [1]. There exist two types of methods for solving such problems: exact and approximate ones [2]. The first group includes integer linear programming [3], dynamic programming [4], the branch and bound method [5], the local elimination algorithm [6], and so on. In this case, the optimal objective function value is calculated without any error, but exact algorithms require large computation times and also a huge memory. Approximate methods such as genetic algorithms [7], ant colony algorithms [8], bee colony algorithms [9], tabu search [10], and many others obtain much faster a heuristic solution but there are usually no estimates of the deviation of the objective function value from the optimal one [11].

In this paper, we describe a general approximation approach which is denoted as metric one [12]. It constructs a solution with a specified maximal absolute error for scheduling problems on parallel machines with the criterion of minimizing maximum lateness. The absolute error of the approximate solution is bounded

© Springer Nature Switzerland AG 2020
N. Olenev et al. (Eds.): OPTIMA 2020, CCIS 1340, pp. 211–223, 2020.
https://doi.org/10.1007/978-3-030-65739-0_16

by a metric function $\rho(A, B)$. The idea of the metric approach is as follows. An instance A of a problem is characterized by a point of the input data. For this problem, we consider all known polynomial and pseudo-polynomial algorithms. Then we introduce some metric. By means of this metric, we find that polynomially or pseudo-polynomially solvable instance with the smallest distance from the given instance. To do this, we compile a system of linear inequalities derived from the initial data. In other words, we construct the projection of the initial point onto a particular instance of a suitable sub-space by the introduced metric.

2   Formulation of the Problem

Scheduling problems were first considered in the middle of the 20th century. For example, the problem of finding an optimal schedule of production items through a sequence of two stages or machines can be found in [13]. The common formulation of the problem is as follows. A set of n jobs $j, j \in N = \{1, \ldots, n\}$, has to be processed on a set of m machines $i, i \in M = \{1, \ldots, m\}$. Preemptions of a job are not allowed. At any time, any machine can process no more than one job.

For each job $j \in N$, a release time r_j, a due date d_j and a processing time p_{ij} for job $j \in N$ on machine $i \in M$ are given with $0 \le p_{ij} \le +\infty$. If $p_{ij} = +\infty$, then job j cannot be processed on machine i. Precedence relations between jobs may be given by an directed acyclic graph $G \subset N \times N$.

A schedule is obtained by partitioning the set N of jobs into subsets N_i of jobs processed on machine $i, i = 1, \ldots, m$. For each set N_i, one has to find the job sequence π_i processed on machine i. Inserted idle times between the processing of jobs are not allowed. This means that, if a job j is assigned to a free machine i and the release time r_j of job j allows to start this job, then it must be started.

Since we consider only regular optimization criteria, the assignment of the jobs to the machines (i.e., the specification of the sets $N_1, \ldots, N_m$) and the job sequences $\pi_1, \ldots, \pi_m$ describe completely a schedule by the set of job sequences: $\pi = \{\pi_1, \ldots, \pi_m\}$. In a semi-active schedule, each job $j \in N$ starts its processing at the earliest possible time: either at the release date r_j, or immediately after the completion of the previous job on this machine, or immediately after the end of a job preceding it according to graph G. Instead of the job sequences, one can equivalently give the starting times s_j of all jobs $j \in N_i, i = 1, \ldots, m$. We denote the set of starting times by $S = \bigcup_{j \in N} s_j$.

Let $Pred(j)$ be the set of all jobs which are a predecessor of job j in the precedence graph G and $(k \to j)_{\pi_i}$ be the jobs scheduled on machine i before job j according to the sequence π_i. Then the starting time of a job $j \in N_i, i = 1, \ldots, m$, in the schedule π is given by

$$s_j(\pi) = \max\left\{r_j, \max_{k \in Pred(j)}(s_k(\pi) + p_{ik}), \max_{(k \to j)_{\pi_i}}(s_k(\pi_i) + p_{ik})\right\}. \qquad (1)$$

Since preemptions of jobs are not allowed, the completion time of job $j \in N_i$ in a schedule π is given by

$$C_j(\pi) = s_j(\pi) + p_{ij}, \; j \in N_i.$$

Definition 1. *The schedule is called feasible, if $r_j \leq s_j(\pi)$ and $C_j(\pi) \leq s_k(\pi)$, for all arcs $(j, k) \in G$ and is denoted by $\bar{\pi}$.*

Remark 1. If a schedule π is known, then the starting times S can be determined; and vice versa, if all starting times S (together with the sets $N_1, \ldots, N_m$) are known, this determines the resulting schedule π.

The optimization criterion is to minimize the maximum lateness:

$$\min_{\pi} \max_{j \in N} \{C_j(\pi) - d_j\}.$$

If $d_j = 0$, for all jobs $j \in N$, the objective turns into the makespan criterion.

Definition 2. *A multi-machine scheduling problem is characterized by*

1. *the job precedence graph G,*
2. *either the makespan criterion or the criterion of minimizing maximum lateness,*
3. *and the parameters: $r = \{r_j\}, d = \{d_j\}, p = \{p_{ij}\}, i \in M, j \in N$.*

This problem is usually denoted by $R|prec, r_j|L_{\max}$ in the classical 3-parameter scheduling problem notation [14].

3   Instances

Let us define the vector of the release times $r = (r_1, \ldots, r_n)$, the $(n \times m)$ matrix p of the processing times p_{ij} and the vector of the due dates $d = (d_1, \ldots, d_n)$ for all machines $i \in M$ and all jobs $j \in N$.

Definition 3. *If there exist fixed parameters r, p and d that form a problem described in Definition 2, we denote it as instance A with the parameters: $r^A = \{r_j^A\}, d^A = \{d_j^A\}, p^A = \{p_{ij}^A\}, j \in N, i \in M$.*

In the rest of this paper, the notation r_j^A means that this value of r_j belongs to the instance A. Thus, the tuple $A = \{G, (r_j^A, p_{ij}^A, d_j^A)\}$, $j \in N, i \in M$ represents all parameters of the instance A.

Definition 4. *Two instances are called isomorphic if they have the same set of optimal schedules.*

For example, instances $A = \{G, (r_j^A, p_{ij}^A, d_j^A)\}$ and $B = \{G, (r_j^B, p_{ij}^A, d_j^A)\}$ are isomorphic if $r_j^B = \alpha r_j^A$, $j \in N$.

Definition 5. *Let the set of parameters r_j, p_{ij}, d_j, $j \in N, i \in M$ of instances be the coordinates of points. We denote the space of such points by $\mathfrak{A}$. Space $\mathfrak{A}$ has $n \cdot (m + 2)$-dimensions.*

Definition 6. *Points of the space $\mathfrak{A}$ are called **P-points**, if there exists a polynomial or pseudo-polynomial algorithm for solving the instance with the corresponding parameters. Let $\widetilde{\mathfrak{A}} \subset \mathfrak{A}$ be the space of all P-points. This space is called* **P-cone.**

For example, for the problem $1|r_j|L_{max}$ the following instance classes are currently known as polynomially solvable: the Jackson class, the Lazarev class and the Hoogeveen class [12]. We can define these classes as follows.

Jackson class of instances: $X \cdot r^T = 0$.

Lazarev class of instances: $X \cdot (d^T - r^T - p^T) \geq 0$, $X \cdot d^T \leq 0$.

Hoogeveen class of instances: $X \cdot (d^T - p^T) \leq (E + \beta) \cdot r^T$, $X \cdot (d^T - p^T) \leq Ed^T$, where E is the identity matrix, and $\beta \in \mathbb{R}$ is some constant value. Hereinafter, we denote the transposed vector by the superscript ^{T}. By X we denote the matrix $n \times n$:

$$X = \begin{pmatrix} 1 & -1 & 0 & \dots & 0 \\ 0 & 1 & -1 & \dots & 0 \\ 0 & 0 & 1 & \dots & 0 \\ & & \dots & & \\ 0 & 0 & \dots & 1 & -1 \\ 0 & 0 & \dots & 0 & 1 \end{pmatrix}.$$

There also exist other polynomially solvable instances that cannot be included into these classes. There are also polynomially solvable classes of instances that are hard to be formalized, for example, the Schrage class of instances.

Next, we will construct a set of instances for multi-machine problems that can be polynomially or pseudo-polynomially solved.

4   Norm and Metric

Consider a set of instances from the problem class $R|prec, r_j|L_{\max}$ with n jobs, m machines and the precedence graph G.

Definition 7. *Two instances are called equal if they have the same scheduling parameters r, p, d and the graph G.*

Definition 8. *Two instances $A = \{G, (r_j^A, p_{ij}^A, d_j^A)| j \in N, i \in M\}$ and $B = \{G, (r_j^B, p_{ij}^B, d_j^B)| j \in N, i \in M\}$ are called equivalent, if*

$$\exists \, d, r : \quad d_j^A = d_j^B + d, \quad r_j^A = r_j^B + r, \quad p_{ij}^A = p_{ij}^B,$$

for all $j \in N, i \in M$.

This definition generates a partition of the sets of instances of the problem into equivalence classes. As a representative of each class, we choose the instance with $r_1 = 0$ and $d_1 = 0$. The resulting class is a $(n \cdot (m + 2) - 2)$-dimensional linear space. Let us denote this space by $\mathfrak{A}_-$. We say that the instance A belongs to the class $\mathfrak{A}_-$ if the condition

$$r_1 = d_1 = 0 \tag{2}$$

is satisfied.

Lemma 1. *Equivalent instances are isomorphic.*

Definition 9. *Next, we consider the following functional on the space of equivalence classes of the instances:*

$$\varphi(A) = \max_{j \in N}\{r_j^A\} - \min_{j \in N}\{r_j^A\} + \max_{j \in N}\{d_j^A\} - \min_{j \in N}\{d_j^A\} + \sum_{j \in N} |p_{ij}^A| \geq 0,$$

for all $A \in \mathfrak{A}_-$.

Theorem 1. *The functional $\varphi(A)$ satisfies the following properties:*

$$\begin{cases} \varphi(A) = 0 \Longleftrightarrow A \equiv 0; \\ \varphi(\alpha A) = \alpha\varphi(A); \\ \varphi(A + B) \leq \varphi(A) + \varphi(B). \end{cases} \tag{3}$$

We have $A = \emptyset$ if $r_j = p_{ij} = d_j = 0$ for all $j \in N$. The first property follows from the definition of the functional $\varphi(A)$. The second one can be checked directly. The third one describes the triangle inequality. Metric functions are separable.

Theorem 2. *[12] Metric functions are separable for polynomially and pseudo-polynomially solvable instances, i.e., for the instance A and the instance B, there exist functionals $\phi(A)$ and $\psi(B)$ as follows:*

$$f(A, B) \leq \phi(A) + \psi(B).$$

Corollary 1. *If for a subproblem $1|r_j|L_{max}$, we define the norm*

$$||A|| = \varphi(A),$$

then the metric functions for the subproblem $1|r_j|L_{max}$ satisfy:

$$\rho(A, B) = \varphi(A) + \psi(B),$$

where $\psi(B) = -\varphi(B)$.

Proof. Consider some points from $\mathfrak{A}$. For convenience, we choose an instance for which $r_1 = 0$ and $d_1 = 0$. Thus, this it represents a $(n \cdot (m + 2) - 2)$-dimensional linear normalized space with the norm

$$||A|| = \varphi(A).$$

It should be noted that this rule leads to the metric defined in (4):

$$\rho(A, B) = ||A - B|| = \varphi(A - B) = \varphi(A) - \varphi(B) = \varphi(A) + \psi(B);$$

$$\psi(B) = -\varphi(B).$$

5   Using Polynomial Subcases

In this paper we intend to introduce the approach for solving the problem $R|prec, r_j|L_{\max}$ in the case when the machines are not identical and each job can have different processing times on different machines.

Definition 10. *Let there be a point (instance) $A \notin \widetilde{\mathfrak{A}}$. Using some metric ρ, we can construct a projection onto the space $\widetilde{\mathfrak{A}}$ with respect to A. The resulting point (instance) $B \in \widetilde{\mathfrak{A}}$ is called the projection of A by the metric ρ.*

Definition 11. *The sub-space $\widetilde{\mathfrak{A}}_\rho^\epsilon(A) \in \widetilde{\mathfrak{A}}$ is called an ϵ–projection of A by the metric ρ if for each of its points $x \in \widetilde{\mathfrak{A}}$, the following inequality is satisfied:*

$$L_{max}^A(\pi^x) - L_{max}^A(\pi^A) \le \epsilon.$$

The metric approach consists of two steps. In the first step, we change the parameters $\{(r_j^A, p_{ij}^A, d_j^A)| j \in N\}$ of the original instance $A = \{G, (r_j^A, p_{ij}^A, d_j^A)\}$, where $j \in N, A \notin \widetilde{\mathfrak{A}}$, so that the projection of A by the metric ρ gives an instance $B = \{G, (r_j^B, p_{ij}^B, d_j^B)| j \in N\}$ in the P-cone. In the next step, we find an optimal schedule π^B for the instance B.

Definition 12. *For two arbitrary instances A and B, we define a **base metric** $\rho(A, B) = \rho_d(A, B) + \rho_r(A, B) + \rho_p(A, B)$, where:*

$$\begin{cases} \rho_d(A, B) = \max\limits_{j \in N}\{d_j^A - d_j^B\} - \min\limits_{j \in N}\{d_j^A - d_j^B\}; \\ \rho_r(A, B) = \max\limits_{j \in N}\{r_j^A - r_j^B\} - \min\limits_{j \in N}\{r_j^A - r_j^B\}; \\ \rho_p(A, B) = \sum\limits_{j \in N}\left(\max\limits_{i \in M}(p_{ij}^A - p_{ij}^B)_+ + \max\limits_{i \in M}(p_{ij}^A - p_{ij}^B)_-\right); \end{cases} \qquad (4)$$

and

$$(x)_+ = \begin{cases} x, & x > 0 \\ 0, & x \le 0 \end{cases},$$

$$(x)_- = \begin{cases} 0, & x \ge 0 \\ -x, & x < 0 \end{cases},$$

i.e., we have $|x| = (x)_+ + (x)_-$.

Theorem 3. *Let the instance B inherit all parameters from the instance A except the values $\{d_j, r_j, p_{ij} \mid j \in N, i \in M\}$, and let $\widetilde{\pi}^B$ be an approximate solution of the instance B satisfying the condition:*

$$L_{\max}^B(\widetilde{\pi}^B) - L_{\max}^B(\pi^B) \le \delta_B. \qquad (5)$$

Then

$$0 \le L_{\max}^A(\widetilde{\pi}^B) - L_{\max}^A(\pi^A) \le \rho(A, B) + \delta_B. \qquad (6)$$

According to Theorem 3, we apply the schedule π^B to the initial instance A. As a result, we obtain the following estimate of the absolute error:

$$0 \leq L^A_{\max}(\pi^B) - L^A_{\max}(\pi^A) \leq \rho(A, B).$$

We consider the P-cone when the parameters of the jobs satisfy the following k linearly independent inequalities:

$$\Omega^B_1 R^B + \Omega^B_2 P^B + \Omega^B_3 D^B \leq H, \tag{7}$$

where $R = r^T$, $P = p^T$, $D = d^T$. Ω^B_3 and Ω^B_2 are $n \times n$ matrices, Ω^B_1 is a vector of m elements, $H = (h_1, \ldots, h_n)^T$ is an n-dimensional vector. For example, for Jackson class of instances $\Omega^B_1 = X$, $\Omega^B_2 = 0$, $\Omega^B_3 = 0$, H is a zero vector.

Then in the class of instances (7), we determine an instance B with minimal distance $\rho(A, B)$ to the original instance A by solving the following linear programming problem:

$$(8)\quad\begin{cases} (x_d - y_d + x_r - y_r) + \sum\limits_{i \in M} \sum\limits_{j \in N} x_{p_{ij}} \longrightarrow \min \\ y_d \leq d^A_j - d^B_j \leq x_d \quad \text{for all } j \in N, \\ y_r \leq r^A_j - r^B_j \leq x_r \quad \text{for all } j \in N, \\ -x_{p_{ij}} \leq p^A_{ij} - p^B_{ij} \leq x_{p_{ij}} \quad \text{for all } i \in M, j \in N, \\ 0 \leq x_{p_{ij}} \quad \text{for all } i \in M, j \in N, \\ \Omega^B_1 R^B + \Omega^B_2 P^B + \Omega^B_3 D^B \leq H. \end{cases}$$

The linear programming problem (8) with $2nm + 2n + 4$ variables $(r^B_j, p^B_{ij}, d^B_j, x_d, y_d, x_r, y_r,$ and $x_{p_{ij}}, i = 1, \ldots, m, j = 1, \ldots, n)$ and $4n+3nm+k$, where k is a number of inequalities in $\Omega^B_1 R^B + \Omega^B_2 P^B + \Omega^B_3 D^B \leq H$. Values y_r and y_d represent a lower bound of differences between a parameter of original instance A and a parameter of projected instance B for all values r and d respectively for all jobs. Similarly, x_r and x_d represent a lower bound of all differences between corresponding parameters. Values $x_{p_{ij}}$, $i \in M, j \in N$ are lower bounds on absolute value of difference between processing time of corresponding job on corresponding machine for instances A and B. These inequalities can sometimes be solved with a polynomial number of operations, depending on the specificity of the constraints of the problem (8). For example, for Hoogeveen and for Lazarev classes the problem can be solved in $O(n^2 \log n)$ [16] and $O(n^3 \log n)$ [15] operations respectively.

In the case when the original problem is not projected onto a P-point, the selected objectives (we can note that trivial P-points usually do not give qualitatively new estimates of the absolute error) or the 'distance' $\rho(A, C)$ to any polynomially solvable instance C is not appropriate. However, if for some instance $B = \{G, (r^B_j, p^B_{ij}, d^B_j) | j \in N\}$, the estimate of the absolute error of the maximum lateness of the approximate schedule $\tilde{\pi}$ is 'acceptable', then the approximate schedule $\tilde{\pi}$ for the original instance $A = \{G, (r^A_j, p^A_{ij}, d^A_j) | j \in N\}$ has a guaranteed absolute error of $\delta^B(\tilde{\pi}) + \rho(A, B)$ from the optimal values of the objective function according to Theorem 3. The value of $\delta^B(\widetilde{\pi}) + \rho(A, B)$ is sometimes significantly less than $\rho(A, C)$ for any polynomially or pseudo-polynomially solvable instance C.

It is an upper bound on the absolute error on the objective function value. In Fig. 1, we show the idea of the metric approach. Let us consider the P-cone which includes all known P-points bounded by a system of linear inequalities (note that, since we consider this sub-space with respect to point zero, it is a cone). In this P-cone, we can find a point B, which corresponds to an instance B with the obtained schedule π^B. Thus, for the instance B, there exists a polynomially or pseudo-polynomially solution algorithm.

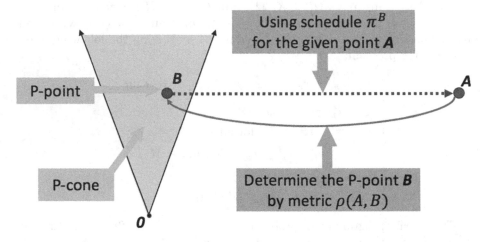

Fig. 1. Geometric illustration of the metric approach

Now we want to find a polynomial or pseudo-polynomial algorithm for an instance A, which does not belong to the P-cone. Then we construct a metric ρ, which characterizes the difference of the two elements. Here the elements are some functions whose parameters allow finding easily schedules for the instances B and A. Thus, the metric looks like $\rho = L^A_{max}(\pi^B) - L^A_{max}(\pi^A)$. That is, the projection of the initial point A from the $n \cdot (m + 2)$-dimension space onto a point of a polynomially solvable sub-space is determined. For any instance (i.e., a point in the $n \cdot (m + 2)$-dimensional space), we know the projection of the initial point A onto the polynomially solvable sub-space. Next, in a first step, we show the efficiency of the metric approach using experiments for the simplified problem $1|r_j|L_{max}$.

6   Generation of Instances for the Problem $1|r_j|L_{max}$

When conducting experimental research, one of the primary issues is the generation of test cases. We show that the set of all instances of the n-dimensional problem can be transformed into a bounded set $\mathfrak{A}^n \subset \mathfrak{A}$ with $3n$ dimensions, where each element represents an infinite number of equivalent instances of the problem. The algorithm presented below generates instances with a uniform distribution over a set of spaces, where the coordinates of each point are the parameters r_j, p_j, d_j, where r_j is a release time, d_j is a due date and p_j is a processing time of job $j \in N$.

Algorithm 1. Algorithm of instance generation on the set $\mathfrak{A}^n$

1: Generate the parameters $\{\hat{r}_j, \hat{p}_j, \hat{d}_j\}$ according to a normal distribution with the location parameter $\nu = 0$ and the variance $\sigma = 1$
2: **for all** $j \in N$ **do**
3:     **if** $\hat{p}_j < 0$ **then**
4:         $p_j := -\hat{p}_j$
5:     **else**
6:         $p_j := \hat{p}_j$
7:     **end if**
8:     $r_j := \hat{r}_j - \min_{j \in N}\{\hat{r}_j\}$
9:     $d_j := \hat{d}_j - \max_{j \in N}\{\hat{d}_j\}$
10: **end for**
11: $\Delta := \sqrt{\sum_{j \in N} r_j^2 + \sum_{j \in N} p_j^2 + \sum_{j \in N} d_j^2}.$
12: **for all** $j \in N$ **do**
13:     $r_j := \frac{r_j}{\Delta};\ p_j := \frac{p_j}{\Delta};\ d_j := \frac{d_j}{\Delta}$
14: **end for**

Next we generate 10000 instances of the problem with the number of jobs up to 50. We do not consider instances with larger number of jobs because the investigated parameters become stable.

7 Efficiency of Using Polynomial Algorithms for the Problem $1|r_j|L_{\max}$

We show how the proposed approach works in special cases of the "easy" sub-problem $1|r_j|L_{\max}$. We generate random instances of the problem $1|r_j|L_{max}$ and build a projection of each instance to the chosen special class. Then, we solve this projected instance with the corresponding polynomial algorithm. Thus, we solve instances of the Lazarev class with the polynomial Lazarev algorithm [11,15], instances of the Hoogeveen class with the polynomial algorithm of Hoogeveen [16], and special instances, which were suggested by Schrage, with the polynomial algorithm of Schrage [17]. For measuring the efficiency we are using the following three indicators:

- μ determines the percentage of instances for which the algorithm found an optimal solution, and it is calculated by the formula:

$$\mu = \frac{K^* \cdot 100\%}{K}. \tag{9}$$

Here K is the total number of generated test instances, K^* is the number of test cases optimally solved by the algorithm.

- The parameters β_{av} and β_{max} determine the average and maximum relative error of the value of the objective function, respectively, found by the scheduling algorithm for the test instance. The error is determined relative to the optimal value of the objective function of the test instance. The parameters β_{av} and β_{max} are calculated using the following formulas:

$$\beta_{av} = \sum_{i=1}^{\bar{K}} \frac{L_{max}(\pi_i) - L_{max}(\pi^*)}{L_{max}(\pi^*)} \cdot 100\%;$$

$$\beta_{max} = \max_{i=1,...,K} \left\{ \frac{L_{max}(\pi_i) - L_{max}(\pi^*)}{L_{max}(\pi^*)} \cdot 100\% \right\}.$$

Here K is the number of generated instances for which the solution obtained by the algorithm was not optimal ($\bar{K} = K - K^*$), π_i and π_i^* are the schedule found by the algorithm and the optimal schedule for the i-th generated instance solved non-optimally.

In our study, we experimentally found the parameters μ, β_{av} and β_{max} for three polynomial algorithms. We studied instances with a number of jobs n ranging from 2 to 50. Test cases were generated according to Algorithm 1 on a $n \cdot (m + 2)$-dimensional unit sphere. For each investigated value of n, 10 000 instances were constructed.

Figure 2 shows the values of the percentage μ, calculated according to (9), of the instances for which each algorithm was able to find an optimal solution, at the x-axis the numbers of jobs are given. We can see that out of 10 000 generated instances, the Hoogeveen algorithm has the largest percentage of optimally solved instances. However, the parameter has a similar dependence for all three algorithms. For a number of jobs $n \leq 6$, the parameter μ decreases with an increase in the number of jobs, and with larger values of n it begins to grow and tends to 100%.

Figures 3 and 4 show the values of the average and maximum relative error, respectively, of the value of the objective function, found by each considered scheduling algorithm. We can see that the relative error of all considered algorithms tends to be smaller with larger values of n.

For example, for dimensions greater than 30, the metric approach finds an optimal solution for more than 99.7% of the instances, and the algorithm of Hoogeveen finds an optimal solution for more than 99.9% of the instances. Of course, there are classes of instances that cannot be optimally solved by these polynomial algorithms. However, if we take into account the whole set of instances of large dimensions, the probability of meeting a "bad" instance is very small. If we consider the parameters β_{av} and β_{max}, then we can see that the average and maximum relative error of the solution obtained by the algorithms decreases with an increase in the number of jobs. Again the algorithm of Hoogeveen shows the best results. The average relative error for this algorithm does not exceed 1% for a dimension of $n_1 = 17$ or more, and the maximum did

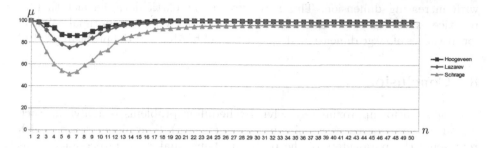

Fig. 2. Percentage μ of optimally solved instances

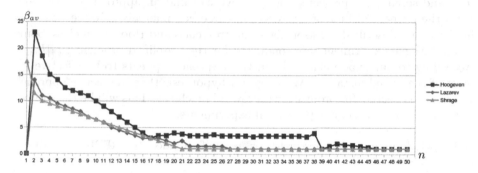

Fig. 3. Average relative error β_{av} of the value of the objective function

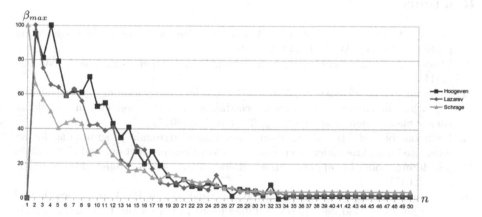

Fig. 4. Maximum relative error β_{max} of the value of the objective function

not exceed 2% in our experiments for the instances of a dimension $n_2 = 23$ or more.

For the metric approach, we obtained $n_1 = 19$ and $n_2 = 36$, respectively. Based on the results of an experimental study, it can be conjectured that the percentage of "difficult" instances relative to the entire set of instances $1|r_j|L_{max}$ decreases with increasing dimension. This may explain, in particular, the fact that in practice exact non-polynomial algorithms can quickly find an optimal solution for instances of large dimension, despite the NP-hardness of the problem [18,19].

8   Conclusion

An approach for approximately solving scheduling problems is analyzed. Each instance is considered as a point in an $n \cdot (m + 2)$-dimensional space, which represents the parameters of the problem. Thus, making a projection of this point to known polynomially solvable areas in the space of problem parameters, and solving the projected instance, we can find an approximate solution. With the purpose of finding a projection, three polynomially solvable classes of instances are described: Jackson class, Lazarev class and Hoogeveen class. Some computational experiments are conducted for the simplified problem $1|r_j|L_{\max}$. According to our experiments, the metric approach appears to be effective for most of the tested instances. We suggest a hypothesis that the metric approach will be also effective for problems with several machines. This will be the subject of further research and computational experiments.

Acknowledgement. This research was partially funded by RFBR and MOST (project 20-58-S52006).

References

1. Lenstra, J., Rinnooy Kan, A.H.G., Brucker, P.: Complexity of machine scheduling problems. Discret. Math. **1**, 343–362 (1977)
2. Brucker, P., Knust, S.: Complex scheduling. Eur. J. Oper. Res. **169**(2), 638–653 (2011)
3. Umetani, S., Fukushima, Y., Morita, H.: A linear programming based heuristic algorithm for charge and discharge scheduling of electric vehicles in a building energy management system. Omega **67**, 115–122 (2017)
4. Rasti-Barzoki, M., Hejazi, S.: Pseudo-polynomial dynamic programming for an integrated due date assignment, resource allocation, production, and distribution scheduling model in supply chain scheduling. Appl. Math. Model. **39**(12), 3280–3289 (2015)
5. Fattahi, P., Hosseini, S., Jolai, F., Tavakkoli-Moghaddam, R.: A branch and bound algorithm for hybrid flow shop scheduling problem with setup time and assembly operations. Appl. Math. Model. **38**(1), 119–134 (2014)
6. Lemtyuzhnikova, D., Leonov, V.: Large-scale problems with quasi-block matrices. J. Comput. Syst. Sci. Int. **58**(4), 571–578 (2019)
7. Werner, F.: A Survey of Genetic Algorithms for Shop Scheduling Problems. Heuristics: Theory and Applications, pp. 161–222. Nova Science Publishers, New York (2013)
8. Zuo, L., Shu, L., Dong, S., Zhu, C., Hara, T.: A multi-objective optimization scheduling method based on the ant colony algorithm in cloud computing. IEEE Access **3**, 2687–2699 (2015)

9. Pan, Q.: An effective co-evolutionary artificial bee colony algorithm for steelmaking-continuous casting scheduling. Eur. J. Oper. Res. **50**(3), 702–714 (2016)
10. Sels, V., Coelho, J., Dias, A., Vanhoucke, M.: Hybrid tabu search and a truncated branch-and-bound for the unrelated parallel machine scheduling problem. Comput. Oper. Res. **53**, 107–117 (2015)
11. Lazarev, A.: Estimation of absolute error in scheduling problems of minimizing the maximum lateness. Dokl. Math. **76**, 572–574 (2007)
12. Lazarev, A.: Scheduling Theory: Methods and Algorithms. ICS RAS, Moscow (2019). (in Russian)
13. Johnson, S.M.: Optimal two- and three-stage production schedules with setup times included. Naval Res. Logist. Q. **1**, 61–68 (1954)
14. Graham, R., Lawler, E., Lenstra, J., Kan, A.: Optimization and approximation in deterministic sequencing and scheduling: a survey. Ann. Discret. Math. **5**, 287–326 (1979)
15. Lazarev, A.: The pareto-optimal set of the NP-hard problem of minimization of the maximum lateness for a single machine. J. Comput. Syst. Sci. Int. **45**(6), 943–949 (2006)
16. Hoogeveen, J.: Minimizing maximum promptness and maximum lateness on a single machine. Math. Oper. Res. **21**, 100–114 (1996)
17. Schrage, L.: Solving resource-constrained network problems by implicit enumeration: non preemptive case. Oper. Res. **18**, 263–278 (1970)
18. Carlier, J.: The one-machine sequencing problem. Eur. J. Oper. Res. **11**(1), 42–47 (1982)
19. Lageweg, B., Lenstra, J., Kan, A.: Minimizing maximum lateness on one machine: computational experience and applications. Stat. Neerlandica **30**, 25–41 (1976)

On Numerical Solving an Equilibrium Problem for a 3D Elastic Body with a Crack Under Coulomb Friction Law

Robert Namm[✉][iD] and Georgiy Tsoy[iD]

Computing Center of the Far Eastern Branch of the Russian Academy of Sciences,
Kim Yu Chen 65, 680000 Khabarovsk, Russia
rnamm@yandex.ru, tsoy.dv@mail.ru
http://www.ccfebras.ru

Abstract. An equilibrium problem for a 3D elastic body with a crack is considered. We assume that nonpenetration boundary conditions and Coulomb friction law are imposed on the crack faces. This leads to the formulation of a problem as a quasi-variational inequality. For solving auxiliary problems with given friction occurring on each outer step of the method of successive approximations we use duality scheme based on the modified Lagrange functional. Computational results illustrating the efficiency of the proposed algorithm are presented.

Keywords: Quasi-variational inequality · Crack problem · Saddle point · Uzawa algorithm · Duality scheme

1 Introduction

In this paper, we consider a boundary value problem describing an equilibrium of an elastic body with a crack. To prevent a mutual penetration between the crack faces, inequality type boundary conditions are imposed at the crack faces [1–6]. Taking into account the Coulomb friction law in the mathematical model of crack problems leads to a quasi-variational inequality [7].

It is well-known that the solution of the quasi-variational inequality can be defined as a fixed-point of a certain mapping and computed by using the method of successive approximations [8–11]. At each step of this method we solve an auxiliary crack problem with given friction. The variational formulation of the auxiliary problem described by a minimization problem of the energy functional on the set of admissible displacements. The computational efficiency of the method of successive approximations depends on the realization of crack problems with given friction. In works [8–10] these auxiliary problems were solved using the classical Lagrange functionals.

This study was supported by the Russian Foundation for Basic Research (Project 20-01-00450 A).

N. Olenev et al. (Eds.): OPTIMA 2020, CCIS 1340, pp. 224–235, 2020.
https://doi.org/10.1007/978-3-030-65739-0_17

In this paper, we use a duality scheme based on the modified Lagrange functionals. The use of modified Lagrange functionals ensures the convergence of the gradient Uzawa algorithm with respect to the dual variable [5,6,11,12]. We apply the finite element method for the numerical implementation of the proposed algorithm and present some computational experiments illustrating its effectiveness.

2 Statement of the Problem

Let $\Omega = (0,1) \times (0,1) \times (0,1)$ be a homogeneous isotropic body with the planar crack $\gamma = (0.25,1) \times (0,1) \times \{0.5\}$, Γ is a Lipschitz boundary of the Ω. Denote by $\Gamma_D = \{0\} \times (0,1) \times (0,1)$ the part of Γ where the body is clamped and by $\Gamma_N^{\pm}$ the parts of Γ where body is loaded by a surface force.

$$\Gamma_N^+ = (0.1,1) \times \{1\} \times (0.6,1), \ \Gamma_N^- = (0.1,1) \times \{0\} \times (0,0.4).$$

In Fig. 1, the parts $\Gamma_N^{\pm}$ are marked and the arrows indicate the directions of the acting forces. Let $\nu = (0,0,1)$ be a unit normal vector on γ. According to the vector ν, denote the positive (upper) face of the crack γ by γ^+ and the negative (lower) face by γ^-. Suppose that $\Omega_\gamma = \Omega \setminus \bar{\gamma}$.

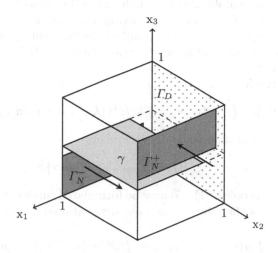

Fig. 1. Domain configuration and loading in the crack problem

Assume that the strains are small. Let us introduce the stress and strain tensors of linear elasticity using the Einstein summation convention

$$\sigma_{ij}(u) = c_{ijkm}\varepsilon_{km}(u), \ \varepsilon_{ij}(u) = \frac{1}{2}\left(u_{i,j} + u_{j,i}\right), \ i,j,k,m = 1,2,3,$$

$$c_{ijkm} = c_{jimk} = c_{kmij}, \ c_{ijkm}\xi_{km}\xi_{ij} \geq c_0|\xi|^2 \ \ \forall \xi_{ij} = \xi_{ji}, \ \ c_0 = const > 0,$$

where $u = (u_1, u_2, u_3)$ is a vector of displacements of the elastic body and lower indices after comma denote the $u_{i,j} = \partial u_i / \partial x_j$.

Let us specify vector-functions of the body and surface forces $f \in L_2(\Omega_\gamma)^3$ and $p \in L_2(\Gamma_N)^3$, respectively. In the domain Ω_γ, we consider the following boundary value problem [1,2]: for the given friction coefficient $\mathscr{F} \geq 0$ find displacements u of the body such that

$$- \operatorname{div} \sigma(u) = f \quad \text{in} \quad \Omega_\gamma, \tag{1}$$

$$u = 0 \quad \text{on} \quad \Gamma_D, \tag{2}$$

$$\sigma(u)n = p \quad \text{on} \quad \Gamma_N, \tag{3}$$

$$[u_\nu] \geq 0, \ [\sigma_\nu(u)] = 0, \ \sigma_\nu(u) \leq 0, \ \sigma_\nu(u)[u_\nu] = 0 \quad \text{on} \quad \gamma, \tag{4}$$

$$[\sigma_\tau(u)] = 0, \ |\sigma_\tau(u)| \leq -\mathscr{F}\sigma_\nu(u), \ \sigma_{\tau i}(u)[u_{\tau i}] + \mathscr{F}\sigma_\nu(u)|[u_\tau]| = 0 \quad \text{on} \quad \gamma, \tag{5}$$

where $\sigma(u)n = (\sigma_1(u), \sigma_2(u), \sigma_3(u)) = (\sigma_{1j}(u)n_j, \sigma_{2j}(u)n_j, \sigma_{3j}(u)n_j)$ and $n = (n_1, n_2, n_3)$ is the unit outward normal vector to Γ; u_ν, u_τ, $\sigma_\nu(u)$, $\sigma_\tau(u)$ are the normal and tangential components of the corresponding vector-functions on γ defined as follows:

$$u_\nu = u_i \nu_i, \ u_{\tau i} = u_i - u_\nu \nu_i, \ i = 1, 2, 3,$$

$$\sigma_\nu(u) = \sigma_{ij}(u)\nu_i \nu_j, \ \sigma_{\tau i}(u) = \sigma_i(u) - \sigma_\nu(u)\nu_i, \ i, j = 1, 2, 3.$$

The boundary value problem (1)–(5) belongs to the class of problems with an unknown contact area. Condition (4) describe a mutual nonpenetration of the crack faces, where $[u_\nu] = u_\nu^+ - u_\nu^-$ is a jump of the function u_ν on γ. The values $u_\nu^\pm \in H^{1/2}(\gamma)$ are the traces of u_ν at the crack faces [2]. At last, conditions (5) define Coulomb friction law.

Let us define the functional space:

$$H^1_\Gamma(\Omega_\gamma) = \left\{ v = (v_1, v_2, v_3) \in H^1(\Omega_\gamma)^3 \mid v = 0 \text{ on } \Gamma_D \right\}$$

and introduce the set of admissible displacements

$$K = \left\{ v \in H^1_\Gamma(\Omega_\gamma) \mid [v_\nu] \geq 0 \text{ on } \gamma \right\}.$$

The equilibrium problem (1)–(5) may be formulated (under the assumption of sufficient regularity of the solution) as a quasi-variational inequality [7,8,13,14]:

$$\text{find } u \in K : \int_{\Omega_\gamma} \sigma(u) : \varepsilon(v - u) \, dx + \int_\gamma \mathscr{F}|\sigma_\nu(u)|(|[v_\tau]| - |[u_\tau]|) ds \geq$$

$$\geq \int_{\Omega_\gamma} f \cdot (v - u) dx + \int_{\Gamma_N} p \cdot (v - u) ds \quad \forall v \in K. \tag{6}$$

Solution of the quasi-variational inequality is described as a fixed-point of mapping $T_\gamma : [H^1(\Omega_\gamma)]^3 \to H^{-1/2}(\gamma)$ defined as follows:

$$(T_\gamma(u), [v]) = a(u, v) - b(v)$$

for all $v \in [H^1(\Omega_\gamma)]^3$ such that $v = 0$ on Γ_D and $[v_\tau] = 0$ on γ [7,8]. Here, $a(u,v)$ is a bilinear form defined on $[H^1(\Omega_\gamma)]^3 \times [H^1(\Omega_\gamma)]^3$:

$$a(u,v) = \int_{\Omega_\gamma} \sigma(u) : \varepsilon(v)\, dx$$

and $b(v)$ is a linear form representing the work of applied forces:

$$b(v) = \int_{\Omega_\gamma} f \cdot v dx + \int_{\Gamma_N} p \cdot v ds.$$

We use the method of successive approximations for solving a quasi-variational inequality (6) [8,9,11]. In our case it reads as follows:

1. **Iteration** $k = 0$. Choose an initial friction force $g^0 \in H^{1/2}(\gamma)$.
2. **Iteration** $k \geq 1$. Define function u^k as a solution of the following variational inequality

$$a(u^k, v - u^k) + \int_\gamma g^{k-1}(|[v_\tau]| - |[u_\tau^k]|)ds \geq b(v - u^k) \quad \forall v \in K. \tag{7}$$

3. Calculate the approximation $g^k = \mathscr{F}|\sigma_\nu(u^k)|$ and repeat Steps 2, 3 until stopping criterion.

Convergence of this method is guaranteed if friction coefficient $\mathscr{F}$ is small enough, but unfortunately it is rather difficult to determine the bounds for $\mathscr{F}$ because they are mesh dependent [8].

The auxiliary variational inequality (7) is called the contact problem with given friction. It can be formulated as the following minimization problem: find $u \in K$ such that

$$J(u) = \inf_{v \in K} J(v), \tag{8}$$

where

$$J(v) = \frac{1}{2}a(v,v) - b(v) + j(v)$$

is the potential energy functional and $j(v)$ is a nondifferentiable functional representing the work of friction forces:

$$j(v) = \int_\gamma g^k |[v_\tau]| ds.$$

It is well-known that problem (8) has a unique solution [1,2]. One of the main differences between 2D and 3D contact problems consists in the definition of the term $|[v_\tau]|$ appearing in $j(v)$. In a two-dimensional case, vector-function v has only one tangential component on γ, so $|[v_\tau]|$ is the absolute value of a vector function, which has only one non-zero component. In three-dimensional case, vector $[v_\tau]$ has the following form:

$$[v_\tau] = ([v_{\tau 1}], [v_{\tau 2}], 0) = ([v_1], [v_2], 0).$$

Therefore, the term $\|[v_\tau]\|$ is defined by $\|[v_\tau]\| = \sqrt{[v_1]^2 + [v_2]^2}$.

In order to smooth functional $j(v)$, we use popular and computationally efficient approximation to $\|[v_\tau]\| \approx \sqrt{[v_1]^2 + [v_2]^2 + \mu}$ (see [17]).

3   Modified Duality Scheme for Solving Problems with Given Friction

To solve problem (8), we define the modified Lagrange functional on the space $H^1_\Gamma(\Omega_\gamma) \times L_2(\gamma)$ (see [5,6]):

$$M(v,l) = J(v) + \frac{1}{2r} \int\limits_\gamma \left([(l - r[v_\nu])^+]^2 - l^2 \right) ds,$$

where $r > 0$ is an arbitrary positive constant.

Definition 1. *A pair* $(v^*, l^*) \in H^1_\Gamma(\Omega_\gamma) \times L_2(\gamma)$ *is called a saddle point of the modified Lagrange functional* $M(v,l)$ *if the following two-sided inequality takes place:*

$$M(v^*, l) \le M(v^*, l^*) \le M(v, l^*)   \forall (v,l) \in H^1_\Gamma(\Omega_\gamma) \times L_2(\gamma).$$

If (v^*, l^*) is the saddle point of $M(v,l)$, then v^* is a solution of the problem (8) and l^* is a solution of the corresponding dual problem:

$$\begin{cases} \underline{M}(l) \to \sup, \\ l \in L_2(\gamma), \end{cases} \tag{9}$$

where

$$\underline{M}(l) = \inf_{v \in H^1_\Gamma(\Omega_\gamma)} M(v, l).$$

In general it is possible to show that a pair $(u, -\sigma_\nu(u)) \in H^1_\Gamma(\Omega_\gamma) \times (H^{-1/2}_\Gamma(\gamma))^+$ is a saddle point of the classical Lagrange functional [8].

Suppose that the solution u to problem (1)–(5) has an additional smoothness. Namely, we assume that

$$([\sigma_{\tau 1}(u)], [\sigma_{\tau 2}(u)], [\sigma_\nu(u)]) \in [L_2(\gamma)]^3.$$

Then it can be shown that a saddle point of modified Lagrange functional coincides with a saddle point of classical Lagrange functional [15]. Moreover, the saddle point has the form $(u, -\sigma_\nu(u))$, and the function $-\sigma_\nu(u)$ is a solution of the dual problem (9) [8–10]. The following statement holds [6].

Theorem 1. *The dual functional $\underline{M}(l)$ is Gateaux differentiable in $L_2(\gamma)$ and its derivative $\nabla \underline{M}(l)$ satisfies a Lipschitz condition with a constant $\dfrac{1}{r}$, that is, the following inequality holds:*

$$\|\nabla \underline{M}(l') - \nabla \underline{M}(l'')\|_{L_2(\gamma)} \le \frac{1}{r} \|l' - l''\|_{L_2(\gamma)} \quad \forall \, l', l'' \in L_2(\gamma).$$

and subdifferential of $\underline{M}(l)$ consists of the single element $\partial \underline{M}(l) = m(l)$

$$m(l) = \max\left(-[u_\nu], -\frac{l}{r}\right) \quad \forall l \in L_2(\gamma).$$

In order to solve (8), we apply the following Uzawa type algorithm [6]:

1. **Iteration** $m = 0$. Choose an arbitrary function $l^0 \in L_2(\gamma)$.
2. **Iteration** $m \ge 1$. Define function u^m as a solution of the following problem

$$u^m = \underset{v \in H^1_\Gamma(\Omega_\gamma)}{\arg\min} \; M(v, l^{m-1}). \tag{10}$$

3. Set

$$l^m = l^{m-1} + r \max\left(-[u^m_\nu], -\frac{l^{m-1}}{r}\right) = (l^{m-1} - r[u^m_\nu])^+. \tag{11}$$

4. Stop or go to step 2.

We obtain the gradient method at the step (11), which has a faster rate of convergence than the gradient projection method applied in the classical duality scheme [5].

4 Computational Results

We assume that the body Ω is made of an elastic isotropic, homogeneous material characterized by Young's modulus $E = 73000$ MPa and Poisson's ratio $\mu = 0.34$ (aluminum). We consider that a volume load is $f = (0, 0, 0)$ and the boundary loading is taken as

$$-\sigma_{22}(u) = \pm 27 \text{ MPa}, \quad -\sigma_{12}(u) = -\sigma_{23}(u) = 0 \quad \text{on } \Gamma^\pm_N.$$

For the numerical solution of the problem, we use the finite element method. We discretize the domain Ω_γ with a crack by a uniform triangulation and apply standard linear finite elements. The number N of nodes of the triangulation and the number of the contact nodes on γ are presented in Table 1 in dependence of the mesh size h.

As stopping criterion for the method of successive approximations, we choose the following condition:

$$\frac{\|g^k - g^{k-1}\|_{L_2(\gamma)}}{\|g^k\|_{L_2(\gamma)}} < 10^{-10}.$$

Table 1. Number of points N for mesh size h.

Mesh size h	0.05	0.025	0.01	0.005
No. of points N	9576	70151	1037876	8150751
No. of contact points N_γ	315	1230	7575	30150

Uzawa algorithm terminates if

$$\max\left(\frac{\|u^m - u^{m-1}\|_{H_\Gamma^1(\Omega_\gamma)}}{\|u^m\|_{H_\Gamma^1(\Omega_\gamma)}}, \frac{\|l^m - l^{m-1}\|_{L_2(\gamma)}}{\|l^m\|_{L_2(\gamma)}}\right) < 10^{-10}.$$

The spaces $H_\Gamma^1(\Omega_\gamma)$ and $L_2(\gamma)$ are approximated by finite-element spaces V_h and L_h, consisting respectively of linear tetrahedron and triangle elements. To solve the finite-dimensional optimization problem obtained after discretization of (8), we use the generalized Newton method [5,16].

Let us give some notations: $\mathbf{A} \in \mathbb{R}^{3N \times 3N}$ is the positive definite stiffness matrix, $\mathbf{b} \in \mathbb{R}^{3N}$ is the vector of the nodal forces, $\boldsymbol{x} = (x_1, \ldots, x_{3N})^T \in \mathbb{R}^{3N}$ is the vector of unknowns, assembling in an appropriate way components of the displacement vector $(u_1(q^i), u_2(q^i), u_3(q^i))^T$ at the triangulation points $q^i \in \mathbb{R}^3$ with $i = \overline{1, N}$, vectors $\boldsymbol{l} \in \mathbb{R}^{N_\gamma}$ and $\boldsymbol{g} \in \mathbb{R}^{N_\gamma}$ collect, respectively, the values of the dual variable (at iteration m) and friction force (at iteration k) at the contact nodes.

Let us introduce the gradient $\boldsymbol{G}(\boldsymbol{x})$ of the finite-dimensional functional obtained after discretization of the problem (8)

$$\boldsymbol{G}(\boldsymbol{x}) = \mathbf{A}\boldsymbol{x} - \mathbf{b} + \boldsymbol{\alpha}(\boldsymbol{x}). \tag{12}$$

For convenience, we denote by $\{i_j^+\}, \{i_j^-\}_{j=1,N_\gamma}$ the numbers of the nodes lying, respectively, on the upper and lower faces of the crack. We assume that jumps $[u_1], [u_2], [u_3]$ across the crack are approximated by differences:

$$[x_j]_1 = (x_{3i_j^+ - 2} - x_{3i_j^- - 2}), \; [x_j]_2 = (x_{3i_j^+ - 1} - x_{3i_j^- - 1}), \; [x_j]_3 = (x_{3i_j^+} - x_{3i_j^-}).$$

Then vector $\boldsymbol{\alpha}(\boldsymbol{x}) \in \mathbb{R}^{3N}$ specified as follows:

$$\boldsymbol{\alpha}_i(\boldsymbol{x}) = \begin{cases} \pm w_j \boldsymbol{g}_j [x_j]_1 / \sqrt{[x_j]_1^2 + [x_j]_2^2 + \mu} & \text{for } i = 3i_j^\pm - 2, \\ \pm w_j \boldsymbol{g}_j [x_j]_2 / \sqrt{[x_j]_1^2 + [x_j]_2^2 + \mu} & \text{for } i = 3i_j^\pm - 1, \\ \mp w_j (l_j - r[x_j]_3)^+ & \text{for } i = 3i_j^\pm, \\ 0 & \text{overwise.} \end{cases}$$

Here $w \in \mathbb{R}^{N_\gamma}$ is the vector of coefficients obtained after discretization of the boundary integral over γ, $\mu = 10^{-14}$ is a small constant. This constant was chosen so that $\sqrt{\mu}$ is smaller than $\max([x_j]_1, [x_j]_2)$ by two orders of magnitude. The generalized Newton method is applied in the following way:

1. Initialize: $x^0 \in \mathbb{R}^{3N}$, $i := 0$.
2. For $i \geq 1$, calculate

$$x^i = x^{i-1} - (\partial G(x^{i-1}))^{-1} G(x^{i-1}).$$ (13)

3. Check stopping criteria

$$\|Ax^i - b + \alpha(x^i)\|_2 < 10^{-12}.$$

Here $\partial G(x)$ is a symmetric sparse matrix defined by:

$$\partial G(x) = A + D(x) \,,$$

where $D(x) \in \mathbb{R}^{3N \times 3N}$ is the generalized Jacobian matrix of $\alpha(x)$ based on a subgradient of its components. The derivative of a ramp function reads as follows:

$$\frac{\partial}{\partial x_{3i_j^+}}(l_j - r[x_j]_3)^+ = \begin{cases} -r, & l_j - r[x_j]_3 > 0, \\ 0, & l_j - r[x_j]_3 \leq 0. \end{cases}$$

Note that, $r > 0$ is an arbitrary positive constant and, in what follows, we put $r = 10^8$. The dependence of the number of the Uzawa algorithm steps on the parameter r was investigated in [5]. The number of Uzawa iterations decreases with an increase of the parameter r.

All numerical experiments were conducted on IBM Power Systems S822LC 8335-GTB server, which is based on two 10-core IBM POWER8 processors with a maximum operating frequency 4.023 GHz and two NVIDIA Tesla P100 GPU accelerators. It should be noted that the generalized Newton method can be efficiently parallelized on GPU using cuSPARSE, cuBLAS libraries. For that, we can substitute (13) by the linear system and solve it using incomplete-Cholesky preconditioned conjugate gradient iterative method [18].

At first, let us give results of the numerical solution with $\mathcal{F} = 0$ (without friction) for various h. The numerical solutions are compared with respect to the potential energy $J(u)$ and the contact zone in the Fig. 2. We can see in Fig. 2b that the contact zones are close to each other and stabilize with decreasing mesh size. In Fig. 2a we observe linear convergence of the values of potential energy functional $J(u)$.

The number of iterations required for convergence of the Uzawa algorithm and the average number of generalized Newton iterations are presented in Table 2. We can see that the number of iterations slightly increases with decreasing h.

Table 2. Number of iterations for mesh size h.

Mesh size h	0.05	0.025	0.01	0.005
No. of iterations (10)	6	7	8	10
No. of iterations (13)	2	3	4	4

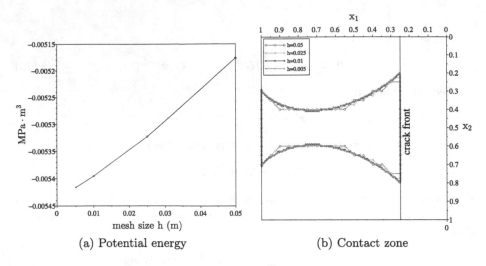

(a) Potential energy   (b) Contact zone

Fig. 2. Solution characteristics by decreasing h.

These facts illustrate the stable behavior of the proposed algorithm with respect to the mesh refinement. Figure 3 shows the values of the jump $[u_3]$ and the dual variable l at the crack. We can observe that $[u_3] \geq 0$, which means that the crack faces do not penetrate into each other. The value of the dual variable is greater than zero at the nodes where the crack faces are closed (in contact). This indicates the presence of the normal stress in these nodes. The presented numerical results for the problem without friction coincide with the results obtained using the primal-dual method in article [3]. A comparison of the classical and modified duality schemes is given in [5].

Let us now consider an example, where the coefficient $\mathscr{F}$ is nonzero, namely $\mathscr{F} = \{0.3, 0.6, 0.9, 1.2\}$. The number of iterations required to successfully termi- nate the method of successive approximations is equal to 3. The average number of iterations of the Uzawa method is equal to 8. The value of the potential energy functional is presented in Table 3.

Table 3. Numerical results for friction coefficient $\mathscr{F}$.

coefficient $\mathscr{F}$	0.3	0.6	0.9	1.2
Energy $J(u)$, $\times 10^{-3}$	-5.34596	-5.29708	-5.24891	-5.20144

Figure 4 shows the computational results for $\mathscr{F} = 1.2$. The graphs show the value of the jumps $[u_1]$, $[u_2]$ (see Fig. 4(a,b)) and friction force $\mathscr{F}|\sigma_\nu(u)|$ (see Fig. 4c) at the crack.

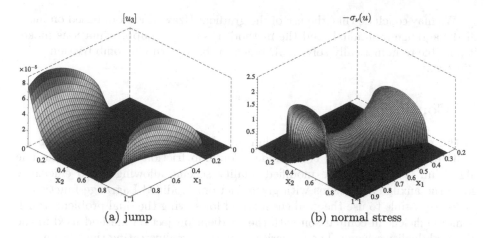

(a) jump

(b) normal stress

Fig. 3. Jump $[u_3]$ and dual variable $l^* = -\sigma_\nu(u)$ at the crack.

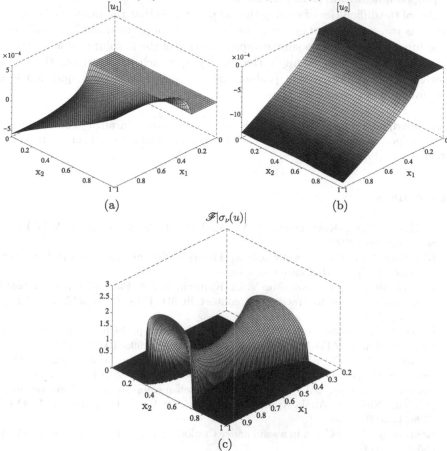

(a)

(b)

(c)

Fig. 4. Jumps $[u_1]$, $[u_2]$ and friction force at the crack.

We may conclude that the use of the gradient Uzawa method, based on modified Lagrange functional, and the method of successive approximations makes it possible to numerically solve a 3D crack problem with Coulomb friction quite effectively.

5   Conclusion

In the paper, the numerical algorithm of solving the equilibrium problem for a 3D elastic body with a crack under Coulomb friction was considered. The algorithm is based on the modified duality scheme allowing us to efficiently solve the auxiliary problems with given friction. Modified Lagrange functionals make it possible to use the gradient method for solving the dual problem, which is more efficient in comparison with the gradient projection method used in the classical duality scheme. The numerical experiments illustrating the efficiency of the proposed algorithm were presented.

One of the difficulties of solving the 3D problem is that the tangential contact stress has two components in each contact node. In order to smooth nondifferentiable functional, we used a simple approximation of the absolute value function. It would be very useful to apply the duality approach to transform the nonsmooth auxiliary minimization problem into a smooth one. This question is a topic for future research.

Acknowledgement(s). This research was supported through computational resources provided by the Shared Facility Center "Data Center of FEB RAS" (Khabarovsk) [19].

References

1. Khludnev, A.M., Kovtunenko, V.A.: Analysis of Crack in Solids. WIT-Press, Southampton (2000)
2. Khludnev, A.M.: Problems of Elasticity Theory in Nonsmooth Domains. Fizmatlit, Moscow (2010). (in Russian)
3. Hintermüller, M., Kovtunenko, V.A., Kunisch, K.: A Papkovich-Neuber-based numerical approach to cracks with contact in 3D. IMA J. Appl. Math. **74**(3), 325–343 (2009)
4. Rudoy, E.M.: Domain decomposition method for crack problems with nonpenetration condition. ESAIM-Math. Model. Numeric. **50**(4), 995–1009 (2016)
5. Namm, R., Tsoy, G.: A modified duality scheme for solving a 3D elastic problem with a crack. Commun. Comput. Inf. Sci. **1090**, 536–547 (2019)
6. Namm, R.V., Tsoy, G.I.: A modified dual scheme for solving an elastic crack problem. Numeric. Anal. Appl. **10**(1), 37–46 (2017). https://doi.org/10.1134/S1995423917010050
7. Kovtunenko, V.A.: Crack in a solid under Coulomb friction law. Appl. Math. **45**(4), 265–290 (2000)
8. Hlaváček, I., Haslinger, Y., Nechas, I., Lovišhek, Y.: Numerical Solution of Variational Inequalities. Springer, New York (1988)

9. Haslinger, J., Kučera, R., Dostál, Z.: An algorithm for the numerical realization of 3D contact problems with Coulomb friction. J. Comput. Appl. Math. **164–165**, 387–408 (2004)

10. Kozubek, T., Vondrák, V.: Massively Parallel Implementation. Scalable Algorithms for Contact Problems. AMM, vol. 36, pp. 319–333. Springer, New York (2016). https://doi.org/10.1007/978-1-4939-6834-3_19

11. Namm, R.V., Tsoi, G.I.: A method of successive approximations for solving the quasi-variational Signorini inequality. Russ. Math. **61**(1), 39–46 (2017). https://doi.org/10.3103/S1066369X17010054

12. Vikhtenko, E.M., Maksimova, N.N., Namm, R.V.: Modified Lagrange functionals to solve the variational and quasivariational inequalities of mechanics. Autom. Remote Control **73**(4), 605–615 (2012)

13. Kikuchi, N., Oden, T.: Contact Problem in Elasticity: A Study of Variational Inequalities and Finite Element Methods. SIAM, Philadelphia (1988)

14. Kravchuk, A.S., Neittaanmäki, P.J.: Variational and quasi-variational inequalities in mechanics. In: Solid Mechanics and its Applications, vol. 147. Springer, Dordrecht (2007)

15. Vikhtenko, E.M., Namm, R.V.: Duality scheme for solving the semicoercive Signorini problem with friction. Comput. Math. Math. Phys. **47**(12), 1938–1951 (2007)

16. Mangasarian, O.L.: A generalized Newton method for absolute value equations. Optim. Lett. **3**(1), 101–108 (2009)

17. Ramirez, C., Sanchez, R., Kreinovich, V., Argaez, M.: $\sqrt{x^2 + \mu}$ is the most computationally efficient smooth approximation to $|x|$: a proof. J. Uncertain Syst. **8**(3), 205–210 (2014)

18. Naumov, M.: Incomplete-LU and Cholesky preconditioned iterative methods using CUSPARSE and CUBLAS. NVIDIA White Paper, pp. 1–16 (2011)

19. Sorokin, A.A., Makogonov, S.V., Korolev, S.P.: The information infrastructure for collective scientific work in the far east of Russia. Sci. Tech. Inform. Process. **44**(4), 302–304 (2017)

On Optimal Selection of Coefficients of a Controller in the Point Stabilization Problem for a Robot-Wheel

Alexander Pesterev[1]([✉])[ID], Yury Morozov[1], and Ivan Matrosov[2]

[1] Institute of Control Sciences, Moscow 117997, Russia
alexanderpesterev.ap@gmail.com
[2] Javad GNSS, Moscow, Russia

Abstract. The point stabilization problem for a robot-wheel is considered. The problem consists in synthesizing control torque in the form of feedback that brings the wheel from an arbitrary initial position on a straight line to a given one, with the control torque and the maximum velocity of wheel motion being constrained. To meet the phase and control constraints, an advanced feedback law in the form of nested saturation functions is suggested. Two of the four coefficients employed in the saturation functions are uniquely determined by the limit value of the control torque and the maximum allowed wheel velocity, while the selection of the other two coefficients can be used to optimize the performance of the controller. In this study, the optimality is meant in the sense that the phase portrait of the closed-loop system is similar to that of a stable node, with the asymptotic rate of decrease of the distance to the target point being as high as possible. The discussion is illustrated by numerical examples.

Keywords: Robot-wheel · Optimal feedback coefficients · Point stabilization problem · Phase and control constraints

1 Introduction

The problem of a wheel rolling on a plane or an uneven terrain is of importance in many practical applications. A rising tide of interest to this classical problem is due to appearance of robotic systems of a new type—ballshaped or spherical robots and robot–wheels—and search for new actuators for such systems [1–5]. The problem of motion control for mobile robots of this type that move owing to displacements of masses (pendulums) inside the shell (wheel) is discussed in many publications (see, for example, [1,3,5,6]). In this paper, we consider the

This work was partially supported by Basic Research Program I.7 "New Developments in Perspective Areas of Energetics, Mechanics and Robotics" of the Presidium of Russian Academy of Sciences and by the Russian Foundation for Basic Research, project no. 18-08-00531.

N. Olenev et al. (Eds.): OPTIMA 2020, CCIS 1340, pp. 236–249, 2020.
https://doi.org/10.1007/978-3-030-65739-0_18

simplest model of a wheel robot assuming that it is driven by a control torque applied to the wheel axis. We do not go into detail of implementation of the actuator assuming only that the control torque is constrained, with the limit value being determined by physical parameters of the robot [1,6]. On the one hand, such a model, in spite of its simplicity, is of interest by itself in the study of advanced control strategies, including optimal ones. On the other hand, this model can be used as a reference one in studying more complicated models, with the solutions obtained for the reference model being taken to be the set of target trajectories for the original system [7].

We set the problem of synthesizing a control law in the form of feedback that brings the wheel from an arbitrary initial position on a straight line to a given one, with the velocity of motion being limited. Moreover, the control torque is also assumed to be constrained. To meet the phase and control constraints, an advanced feedback law in the form of nested saturation functions depending on four coefficients is suggested. Feedback laws of this type were studied in [8,9]. The basic advantage of such laws is that they ensure global stability of the closed-loop system and guarantee the fulfilment of the phase and control constraints under appropriate choice of feedback coefficients.

Two of the four feedback coefficients are uniquely determined by the limit value of the control torque and the maximum allowed wheel velocity, while the selection of the other two coefficients can be used to optimize the performance of the controller. In this study, we use the same optimality criterion as in [10], where the selection of feedback coefficients of a saturated linearizing feedback for a wheeled robot with constrained control resource was discussed. The optimality is meant in the sense that the phase portrait of the nonlinear closed-loop system is similar to that of a linear system with a stable node, with the asymptotic rate of approaching the target point being as high as possible.

2   Problem Statement

We consider a wheel rolling on a plane along a straight line (Fig. 1). The dynamics of the wheel are given by

$$M\ddot{x} = R, \; Mr^2\ddot{\theta} = rR - f\dot{\theta} - U,$$

where M and r are mass and radius of the wheel, x is the coordinate of the wheel center, θ is the rotation angle, R is the reaction force, f is the viscous friction coefficient, and U is the control torque.

We assume that the wheel rolls without slipping; i.e., the condition

$$\dot{x} + r\dot{\theta} = 0 \tag{1}$$

holds. In the point stabilization problem, it is required to synthesize a control law U in the form of feedback that brings the wheel to a given target point on the line. Without loss of generality, we set the target point to be at the origin.

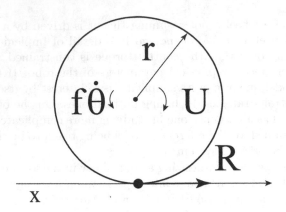

Fig. 1. Schematic of the robot-wheel.

The control torque U is assumed to be limited, and we also assume that the velocity of the wheel center cannot exceed a prescribed value:

$$|U| \le U_{max}, \; |\dot{x}| \le V_{max}. \tag{2}$$

First, we reduce the order of the system and exclude the reaction force. By using the nonslipping condition (1), we get

$$\mu\ddot{x} = -\frac{f\dot{x}}{r^2} + \frac{U}{r}, \tag{3}$$

where $\mu = 2M$. Then, we arrive at the following statement: Find feedback $U = U(x, \dot{x})$ that stabilizes solution of Eq. (3) at zero subject to the phase and control constraints (2).

3 Synthesis of Feedback

Consider the feedback in the form of nested saturators

$$U(x, \dot{x}) = -rk_4 \text{Sat}(k_3(\dot{x} + k_2 \text{Sat}(k_1 x))) + \frac{f\dot{x}}{r}, \tag{4}$$

where $\text{Sat}(x)$ is the saturation function defined by the conditions $\text{Sat}(x) = x$ for $|x| \le 1$ and $\text{Sat}(x) = \text{sign}(x)$ for $|x| > 1$ and $k_i > 0$, $i = 1, 2, 3, 4$, are positive coefficients. Substituting (4) into (3), we get the following equation governing the closed-loop system:

$$\mu\ddot{x} = -k_4 \text{Sat}(k_3(\dot{x} + k_2 \text{Sat}(k_1 x))). \tag{5}$$

It is not difficult to prove that, for any initial condition, solution of Eq. (5) asymptotically tends to zero.

Lemma 1. *Let*

$$U_{max} - fV_{max}/r > 0, \tag{6}$$

and let

$$k_2 = V_{max}, \ k_4 = \frac{U_{max} - fV_{max}/r}{r}. \tag{7}$$

Then, if $|\dot{x}(0)| \le V_{max}$, constraints (2) hold for any positive k_1 and k_3.

Proof. First, let us prove that, if $k_2 = V_{max}$, then the phase constraint holds. In the phase plane, consider the line

$$\sigma(x, \dot{x}) = 0, \ \sigma(x, \dot{x}) = \dot{x} + V_{max}\text{Sat}(k_1 x). \tag{8}$$

Clearly, the line consists of three straight segments $\dot{x} = V_{max}$ when $x \le -1/k_1$, $\dot{x} = -k_1 x$ when $-1/k_1 < x < 1/k_1$, and $\dot{x} = -V_{max}$ if $x \ge 1/k_1$, which all lie in the strip $|\dot{x}| \le V_{max}$. Consider a trajectory of (5) beginning in the strip. Let, for definiteness, the initial point lie under line (8) (the case where the initial point lies from the other side of the line is analyzed similarly). Since the right-hand side of (5) is positive in this region, $\ddot{x} > 0$ and $\dot{x}$ grows until it reaches line (8), where the acceleration vanishes. Obviously, the trajectory cannot intersect the line in the region where $x < -1/k_1$ (the acceleration from the other side is negative!) and asymptotically approaches it such that $\dot{x} < V_{max}$. The trajectory can intersect the line in the region $-1/k_1 < x < 0$, where $\dot{x} < V_{max}$. Since, on the line (8), $\ddot{x} = 0$, the velocity reaches maximum at the intersection point and cannot leave the strip. Finally, if $x(0) > 0$, then $\dot{x}(0) < 0$, and there exist two possibilities: either the trajectory goes to the origin with $x(t)$ staying positive or it intersects the $\dot{x}$-axis, which brings us at the variant considered previously. Hence, we proved that, if $k_2 = V_{max}$ and $|\dot{x}(0)| \le V_{max}$, then, for all $t > 0$, $|\dot{x}(t)| \le V_{max}$ for any positive k_1, k_3, and k_4.

Now, let k_4 be defined as in (7). From (4), (6), and (7) it follows that

$$|U| \le rk_4 + \frac{f|\dot{x}|}{r} \le rk_4 + \frac{fV_{max}}{r} = U_{max} - \frac{fV_{max}}{r} + \frac{fV_{max}}{r} = U_{max};$$

i.e., under the assumptions of the lemma, the control constraint also holds.

As can be seen, the selection of the coefficients k_2 and k_4 ensures the fulfillment of the phase and control constraints (2), which hold for any positive values of the two other coefficients. Hence, the coefficients k_1 and k_3 can be used to optimize the performance of the controller, which is discussed in the remainder of the paper.

The analysis is greatly simplified if we write all equations in terms of dimensionless variables and parameters.

4 Dimensionless Model

Let us introduce the dimensionless time and coordinate:

$$\tilde{t} = tV_{max}/r, \ \tilde{x} = x/r. \tag{9}$$

The derivatives of x in terms of the new variables are written as

$$\dot{x} = V_{max}\frac{d\tilde{x}}{d\tilde{t}}, \quad \ddot{x} = \frac{V_{max}^2}{r}\frac{d^2\tilde{x}}{d\tilde{t}^2}.$$

Substituting these into (3)–(5), introducing dimensionless parameters,

$$\tilde{U} = U/U_{max}, \quad \tilde{\mu} = \frac{\mu V_{max}^2}{U_{max}}, \quad \tilde{f} = \frac{f V_{max}}{r U_{max}},$$

$$\tilde{k}_1 = rk_1, \quad \tilde{k}_2 = k_2/V_{max}, \quad \tilde{k}_3 = k_3 V_{max}, \quad \tilde{k}_4 = \frac{r}{V_{max}^2}k_4,$$

and using the dot notation for the derivatives with respect to the new time, we get Eq. (3) in the dimensionless form:

$$\tilde{\mu}\ddot{\tilde{x}} = -\tilde{f}\dot{\tilde{x}} + \tilde{U}. \tag{10}$$

Constraints (2) take the form

$$|\tilde{U}| \leq 1, \quad |\dot{\tilde{x}}| \leq 1. \tag{11}$$

As can be seen, the dimensionless model depends on only two parameters, $\tilde{\mu}$ and $\tilde{f}$, compared to the five—μ, f, r, U_{max}, and V_{max}—parameters in the dimensional one. Assumption (6) in Lemma 1 turns to the inequality

$$0 \leq \tilde{f} < 1. \tag{12}$$

From Lemma 1, it follows that

$$\tilde{k}_2 = 1, \quad \tilde{k}_4 = 1 - \tilde{f}, \tag{13}$$

the stabilizing control ensuring the fulfillment of constraints (11) is written as

$$\tilde{U}(\tilde{x}, \dot{\tilde{x}}) = -(1 - \tilde{f})\mathrm{Sat}(\tilde{k}_3(\dot{\tilde{x}} + \mathrm{Sat}(\tilde{k}_1\tilde{x}))) + \tilde{f}\dot{\tilde{x}}, \tag{14}$$

and Eq. (5) governing the closed-loop system takes the form

$$\tilde{\mu}\ddot{\tilde{x}} = -(1 - \tilde{f})\mathrm{Sat}(\tilde{k}_3(\dot{\tilde{x}} + \mathrm{Sat}(\tilde{k}_1\tilde{x}))). \tag{15}$$

The coefficients $\tilde{k}_1$ and $\tilde{k}_3$ may take arbitrary positive values and will be further used to optimize the performance of the controller.

To simplify subsequent calculations and formulas, we will use the same notation (without tilde) as in the dimensional case to denote dimensionless quantities and parameters.

5   Optimization Problem Statement

It is easy to see that the closed-loop system (15) is piecewise linear. In the intersection of the sets $|x| \leq 1/k_1$ and $|\dot{x} + k_1x| \leq 1/k_3$, which includes the origin, it takes the form

$$\mu\ddot{x} + k_3k_4\dot{x} + k_1k_3k_4x = 0, \quad k_4 = 1 - f. \tag{16}$$

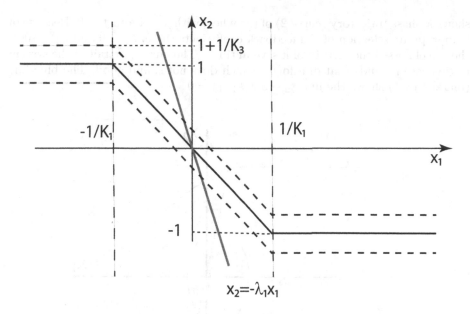

Fig. 2. Partition of the phase plane for system (15).

Let us introduce the notation $x_1 = x$ and $x_2 = \dot{x}$. In Fig. 2, the boundaries of these two sets are depicted by two vertical dashed lines $x_1 = \pm 1/k_1$ and by two inclined dashed lines from the both sides of the line

$$x_2 + \text{Sat}(k_1 x_1) = 0 \tag{17}$$

(depicted by the bold solid line), where the right-hand side of (15) vanishes.

Depending on the values of k_1 and k_3, the origin is either a stable node or focus. In the latter case, the wheel will approach the equilibrium performing oscillations around the target point with a decreasing amplitude, which seems to be undesirable. In the case of a node, the wheel approaches the target point either monotonically or has only one overshooting, when it misses the origin once and then monotonically approaches it from the other side. From (16), it is easily derived that, in order that the origin be a node, the coefficients must satisfy the inequality

$$k_3 k_4 \geq 4 k_1 \mu. \tag{18}$$

In what follows, we assume that inequality (18) holds.

As proved in Lemma 1, any trajectory of the equation beginning in the strip $|x_2| \leq 1$ never leaves it, i.e., the strip is an invariant set of the system, and intersects line (17) only when $|x_1| \leq 1/k_1$ being governed by the linear Eq. (16). Further, the system can stay in the considered set until it reaches the equilibrium or leave it depending on the values of the coefficients k_1 and k_3. Switching from one linear mode to another and vice versa can happen several times, which may result in large overshooting, like in the case presented in Fig. 3. The figure shows a phase trajectory (curve 2) of the wheel with $\mu = 1$ and $f = 0$. Because of inappropriate selection of the feedback coefficients (here, $k_1 = 9$ and $k_3 = 100$), the wheel missed the target point several times. The phase portrait of the system in this case reminds that of a focus, which does not sound good. The blue line (marked by 1) shows the line $x_2 + \mathrm{Sat}(k_1 x_1) = 0$,

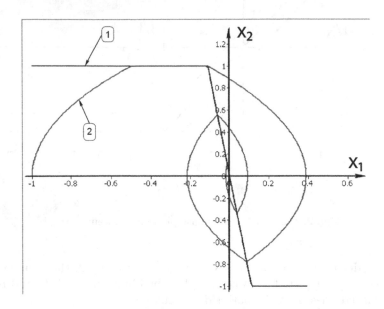

Fig. 3. An example of inappropriate selection of feedback coefficients in (15). (Color figure online)

To get an idea of how the feedback coefficients affect system behavior, we first consider the phase portrait of the linear system (16), which governs the closed-loop system behavior in the neighborhood of the origin.

Let ν_1 and ν_2 be roots of the characteristic equation of (16). Introducing the notation $\lambda_1 = -\nu_1$ and $\lambda_2 = -\nu_2$, we rewrite Eq. (16) in the form

$$\ddot{x} + (\lambda_1 + \lambda_2)\dot{x} + \lambda_1 \lambda_2 x = 0. \tag{19}$$

By virtue of (18), λ_1 and λ_2 are positive real numbers. A typical phase portrait of a system with a stable node is shown in Fig. 4. Here, $\lambda_1 = 2.6$, and $\lambda_2 = 6.3$. The system has two eigenvectors collinear to the straight lines $x_2 = -\lambda_1 x_1$ and $x_2 = -\lambda_2 x_1$. Any system trajectory, except those beginning at the points on the straight line corresponding to the larger eigenvalue (λ_2), touches at the origin the asymptote

$$x_2 + \lambda_1 x_1 = 0 \tag{20}$$

where

$$\lambda_1 = \frac{k_3}{2\mu}\left(1 - \sqrt{1 - \frac{4\mu k_1}{k_3}}\right). \tag{21}$$

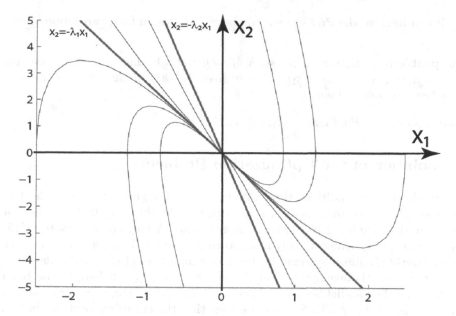

Fig. 4. Phase portrait of a linear system with stable node at the origin.

The asymptote divides the phase plane into two half-planes A_- (below the asymptote) and A_+ (above the asymptote), where the left-hand side of (20) is less or greater than zero, respectively. Clearly, A_- and A_+ are invariant sets of system (19), i.e., any trajectory completely lies in one of these half-planes and may intersect the line $x_1 = 0$ not more than once. The deviation decreases exponentially with the exponent λ_1.

Intuitively, speaking of desirable behavior, we want to have fast asymptotic convergence to the origin in the time domain and the phase portrait of the nonlinear system to look like that of a linear system with a node, when any trajectory approaches the origin monotonically, or has at most one overshooting. The latter property can formally be defined as follows.

Definition 1. *We will say that the phase portrait of the nonlinear system (15) is of the* node-like *type if there exists a straight line, further referred to as an asymptote, that divides the strip* $|x_2| \leq 1$ *into two invariant sets such that any phase trajectory of the system beginning in the strip completely lies in one of the invariant sets.*

Clearly, the asymptote of the nonlinear system (15) in this definition, if exists, must coincide with asymptote (20) of the linear system (16).

Now, the problem we are going to solve can be stated as follows.

Problem. *Determine feedback coefficients* k_1 *and* k_3 *for which the asymptotic rate of approaching the target point is maximal under the condition that the phase portrait of system (15) is of the node-like type.*

It can be seen, the Problem stated above includes, in fact, two subproblems.

Subproblem 1. *Determine the set K of coefficients k_1 and k_3 for which the line $x_2 = -\lambda_1(k_1, k_3)x_1$, where $\lambda_1(k_1, k_3)$ is given by (21), divides the set $|x_2| \leq 1$ into two invariant subsets.*

Subproblem 2. *Find $\max \lambda_1(k_1, k_3)$ on the set K.*

6   Solution of the Optimization Problem

Let us determine conditions the fulfillment of which guarantees that the trajectories of the nonlinear system (15) beginning in the strip $|x_2| \leq 1$ do not intersect the asymptote of the linear system (16). A trajectory of system (15) can intersect the asymptote only in a saturation region. Thus, in order that the asymptote of the linear system (16) be the asymptote of (15), it will suffice that the segment of the asymptote lying in the strip $|x_2| \leq 1$ belong to the band confined by the parallel lines $x_2 + k_1 x_1 = -1/k_3$ and $x_2 + k_1 x_1 = 1/k_3$ (dashed lines in Fig. 2). From Fig. 2, it can be seen that the entire segment lies in this region if and only if the point of the intersection of the asymptote and the line $x_2 = 1$ belongs to it, i.e., the coordinates of the intersection point $(-1/\lambda_1, 1)$ satisfy the inequality $|x_2 + k_1 x_1| \leq 1/k_3$. Substituting them into the inequality, we get

$$1 - \frac{k_1}{\lambda_1} \leq \frac{1}{k_3}. \tag{22}$$

Comparing two representations of the same linear equation, (16) and (19), we find expressions for k_1 and k_3 in terms of λ_1 and λ_2:

$$k_1 = \frac{\lambda_1 \lambda_2}{\lambda_1 + \lambda_2}, \quad k_3 = \frac{\mu(\lambda_1 + \lambda_2)}{k_4}. \tag{23}$$

Substituting them into (22) for k_1 and k_3, we obtain

$$\frac{\lambda_1}{\lambda_1 + \lambda_2} \leq \frac{k_4}{\mu(\lambda_1 + \lambda_2)},$$

from which it follows that, for any λ_2, the required segment of the asymptote completely lies in the linearity region when $\lambda_1 \leq k_4/\mu$. Hence, we have proved the following

Lemma 2. *Let λ_1 and λ_2 be positive numbers such that $0 < \lambda_1 \leq k_4/\mu$ and $\lambda_1 \leq \lambda_2$. Let k_1 and k_3 be given by (23). Then, the phase portrait of the nonlinear system (15) is of the node-like type, with the asymptote being given by $x_2 = -\lambda_1 x_1$.*

From Lemma 2, it follows that the set $\mathcal{K}$ can be defined as the two-parameter family (23), where $0 < \lambda_1 \leq k_4/\mu$ and $\lambda_2 \geq \lambda_1$.

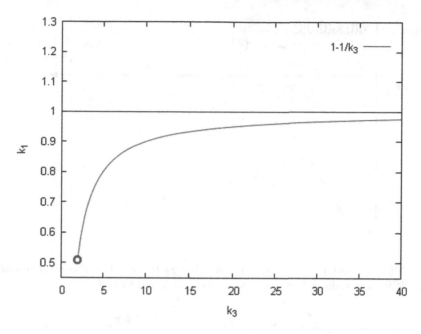

Fig. 5. The set of coefficients k_1 and k_3 ensuring the greatest rate of the asymptotic convergence while preserving the node-like phase portrait of the nonlinear system (15) with $\mu = 1$ and $f = 0$.

The highest convergence rate is obviously achieved when $\lambda_1 = k_4/\mu$ and does not depend on λ_2. Taking into account that $k_4 = 1 - f$, substituting $\lambda_1 = (1 - f)/\mu$ into (23) and taking k_3 as a parameter, we arrive at the solution to the above-stated Problem given by the following

Theorem 1. *The partition of the strip* $|x_2| \leq 1$ *into two invariant subsets with the greatest exponential rate* $\lambda_1 = (1 - f)/\mu$ *of the deviation* x *decrease occurs for the family of the coefficients*

$$k_3 \geq 2, \quad k_1 = \frac{1 - f}{\mu}\left(1 - \frac{1}{k_3}\right). \tag{24}$$

Theorem 1 implies that there exist an infinite number of pairs of the parameters k_1 and k_3 given by (24) for which we have the same partition of the strip $|x_2| \leq 1$ and the same asymptotic rate of the deviation decrease. For any pair from this family, system (16) has the same lower eigenvalue $\lambda_1 = k_4/\mu$, and the same asymptote $x_2 = -\lambda_1 x_1$ that divides the strip $|x_2| \leq 1$ into two invariant sets. Any pair (k_3, k_1) from this family lies on the curve defined by (24). Figure 5 shows the plot of this curve for the system with $\mu = 1$ and $f = 0$.

Let us see how the value of k_3 affects the behavior of the closed-loop system. The second eigenvalue $\lambda_2 = \lambda_1(k_3 - 1)$ is equal to λ_1 when $k_3 = 2$ and goes to infinity when $k_3 \to \infty$. Since $k_1 \to \lambda_1$ and $1/k_3 \to 0$ as $k_3 \to \infty$, the slope of the inclined segment of line (17) grows and the width of the linearity region reduces.

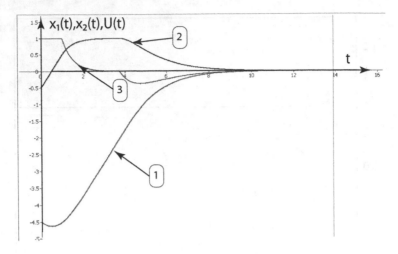

Fig. 6. Plots of deviation x_1 (curve 1), velocity x_2 (curve 2), and control torque U (curve 3) for the optimal $\lambda_1 = 1$ and $\gamma = 1$ ($k_1 = 0.5$ and $k_3 = 2$).

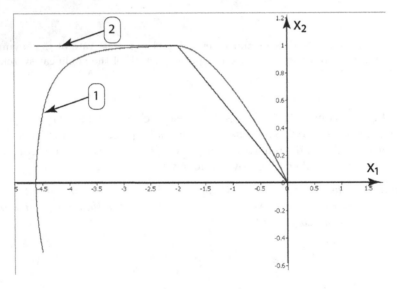

Fig. 7. Phase trajectory (curve 1) and line (17) (curve 2) for the optimal $\lambda_1 = 1$ and $\gamma = 1$ ($k_1 = 0.5$ and $k_3 = 2$).

In the limit, the inclined segment merges with the asymptote, and the linearity region reduces to the line (17) with $k_1 = \lambda_1$.

The last four figures show results of numerical experiments with the wheel with $\mu = 1$ and $f = 0$ and optimal $\lambda_1 = 1$. Figures 6 and 8 present plots of deviation x_1 (curve 1), velocity x_2 (curve 2), and control torque U (curve 3) in the time domain for the pairs ($k_3 = 2$, $k_1 = 0.5$) and ($k_3 = 100$, $k_1 = 0.99$),

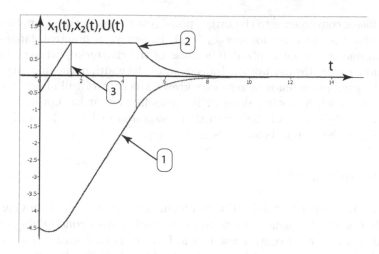

Fig. 8. Plots of deviation x_1 (curve 1), velocity x_2 (curve 2), and control torque U (curve 3) for the optimal $\lambda_1 = 1$ and $\gamma = 100$ ($k_1 \approx 0.99$ and $k_3 = 101$).

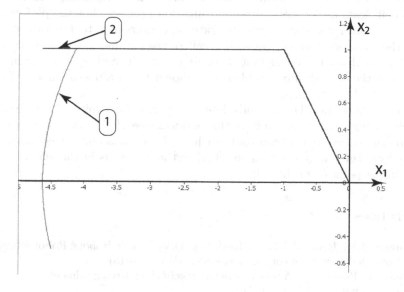

Fig. 9. Phase trajectory (curve 1) and line (17) (curve 2) for the optimal $\lambda_1 = 1$ and $\gamma = 100$ ($k_1 \approx 0.99$ and $k_3 = 101$).

respectively. The corresponding phase trajectories are shown in Figs. 7 and 9, respectively. As can be seen, the increase in k_3 results in a more aggressive control. The wheel moves with the maximum acceleration ($U = \pm 1$) until it reaches the maximum velocity, after which it slides along line (17), which is clearly seen in Fig. 9, where, after the system reached maximum velocity, two curves coincide. Theoretically, greater values of k_3 are more preferable since they result in faster convergence to the target point (even though the asymptotic rate of convergence is the same for any $k_3 \geq 2$). However, in practice, large values of this parameter are not applicable because of the chattering arising when the system moves along the asymptote. Indeed, since the width of the linearity region around the asymptote tends to zero as γ grows, the control will alternately take limit values ± 1 when moving along it. It seems likely that the optimal solution should be a hybrid control similar to that suggested in [10]. The study of the effect of k_3 on the system behavior is underway.

7   Conclusions

In the paper, the point stabilization problem for a robot–wheel moving along a straight line on the plane subject to phase and control constraints has been discussed. An advanced control law in the form of nested saturators has been suggested. The system closed by this feedback is shown to be asymptotically stable in the whole and satisfies both the control and phase constraints for any positive values of the two feedback coefficients. It has been suggested to select these coefficients as solutions of an optimization problem. For the performance index, we considered the asymptotic rate of convergence to the target point under the condition that the phase portrait of the nonlinear closed-loop system is similar to that of a linear system having a stable node at the origin. The solution of the optimization problem was shown to be a one-parameter family of the feedback coefficients.

In the future, we plan to study how the particular choice of the feedback coefficients from this family affects the performance of the controller. We also plan to synthesize a hybrid control law where the selection of the feedback coefficients from the family depends on whether the system is in the neighborhood of the target point or far from it.

References

1. Borisov, A.V., Pavlovskii, D.V., Treshchev, D.V.: Mobile Robots: Robot-wheel and Robot-ball. Institute of computer studies, Izhevsk (2013)
2. Chase, R., Pandya, A.: A review of active mechanical driving principles of spherical robots. Robotics **1**(1), 3–23 (2012)
3. Kilin, A.A., Pivovarova, E.N., Ivanova, T.B.: Spherical robot of combined type. Dyn. Control. Regul. Chaotic Dyn. **20**(6), 716–728 (2015)
4. Ylikorpi, T.J., Forsman, P.J., Halme, A.J.: Dynamic obstacle overcoming capability of pendulum driven ball-shaped robots. In: Proceedings of the 17th IASTED International Conference Robotics Applications, pp. 329–338 (2014)
5. Bai, Y., Svinin, M., Yamamoto, M.: Dynamics-based motion planning for a pendulum-actuated spherical rolling robot. Regular Chaotic Dyn. **23**(4), 372–388 (2018). https://doi.org/10.1134/S1560354718040020
6. Chernous'ko, F.L., Ananievski, I.M., Reshmin, S.A.: Control of Nonlinear Dynamical Systems: Methods and Applications. Springer, Heidelberg (2008). https://doi.org/10.1007/978-3-540-70784-4

7. Matrosov, I.V., Morozov, Yu.V. Pesterev, A.V.: Control of the robot-wheel with a pendulum. In: Proceedings of the 2020 International Conference "Stability and Oscillations of Nonlinear Control Systems" (Pyatntitskiy's Conference), pp. 1–4. IEEE (2020). https://doi.org/10.1109/STAB49150.2020.9140489
8. Saberi, A., Lin, Z., Teel, A.: Control of linear systems with saturating actuators. lIEEE Trans. Autom. Control 41(3), 368–378 (1996)
9. Olfati-Saber, R.: Global stabilization of a flat underactuated system: the inertia wheel pendulum. In: IEEE Conference on Decision and Control, vol. 4, pp. 3764–3765 (2001)
10. Pesterev, A.V.: Synthesis of a stabilizing feedback for a wheeled robot with constrained control resource. Autom. Remote Control 77(4), 578–593 (2016). https://doi.org/10.1134/S0005117916040044

Author Index

Printed in the United States
By Bookmasters